PLASTICS
HISTORY
U.S.A.

*Many of the pioneer molding methods and the machinery were devel-
oped by the molders of rubber, as shown in this 1890 picture. (Courtesy,
B. F. Goodrich Chemical Company, Cleveland, Ohio.)*

PLASTICS HISTORY U.S.A.

J. HARRY DUBOIS

Sponsored by
Plastics World, a Cahners Publication

CAHNERS BOOKS, *Division of Cahners Publishing Co., Inc.*
89 Franklin Street, Boston, Massachusetts 02110

International Standard Book Number: 0–8436–1203–7
Library of Congress Catalog Card Number: 79–156480
Copyright © 1972 by Cahners Publishing Company, Inc.
Printed in the United States of America.
Halliday Lithograph Corporation, Hanover, Massachusetts, U.S.A.

CARGOES*

By John Masefield

Quinquireme of Nineveh from distant Ophir,
Rowing home to haven in sunny Palestine,
With a cargo of ivory,
And apes and peacocks,
Sandalwood, cedarwood, and sweet white wine.

Stately Spanish galleon coming from the Isthmus,
Dipping through the Tropics by the palm-green shores
With a cargo of diamonds,
Emeralds, amethysts,
Topazes, and cinnamon, and gold moidores.

Dirty British coaster with a salt-caked smoke stack,
Butting through the Channel in the mad March days,
With a cargo of Tyne coal,
Road-rails, pig-lead,
Firewood, iron-ware, and cheap tin trays.

L'Envoi by J. Harry DuBois:

Silvery eagles span the ocean, cloud borne and free,
Spreading wings of glass and polyester,
With a cargo of people,
Computers and jewels,
Perishable foods in polyolefins.

Contents

Preface

Two generations have seen the plastics industry grow from a handmade-comb-and-novelty business into an industrial giant that is destined to dominate all of the basic materials. In ten brief years, the volume of plastics is expected to equal that of the metals. By the year 2000, the volume of plastics will exceed that of all basic materials.

There is much to be learned for contemporary practice from the painful steps and the problems that directed the actions in the past. It is difficult now to understand why thermosetting molders were so slow in envisioning the benefits from injection molding. At times there were violent reactions to changes within the industry that are ridiculous in retrospect. Greatest progress was made repeatedly by those who were not prejudiced by the past and were willing to take advantage immediately of new materials and developments. In retrospect, I would change the old adage to read: "Be not the last to try the new nor last to drop the old."

This review is concerned with the men, their machines and materials. The business problems are common in all trades. In the beginning there were no plants, no processes, no proven techniques, no accumulated experience, no customers, and no money. During these tender years, an infinite number of obstacles had to be overcome. There were deep-rooted prejudices among the potential customers. Early salesmen had great difficulty in gaining an audience for this radical industry. Bankers were afraid to lend money to the pioneers. During World War I, the government took over the basic ingredients so that substitutes had to be developed. No textbooks were available; no industry magazines had been started; every molder guarded his techniques, kept his shop locked to salesmen, and was sure his competitor was a bumbling amateur. Materials makers made materials and knew very little about their processing—except that which they learned from their customers.

Looking backward from 1971, it is obvious that we would have made greater progress by knowing the growth steps and the factors that pointed the way to profitable operations. Progress was slowed down by prejudices inherited from the immediate past and by historical ignorance.

Plastics History — U.S.A. will be very helpful to plastics engineers, machinery builders, research workers, patent engineers, educators, students, salesmen, copy writers, and investment specialists. There is here much that will be nostalgic for the present and the past generations of plastics people. Much inspiration and many sources of information came from the **Plastics Pioneers Association** and their historical tapes. The bibliography will be most helpful to those who wish to explore the past in greater detail.

> J. H. DuBois
> Craftsman Farms
> Morris Plains,
> New Jersey

PLASTICS
HISTORY
U.S.A.

1

The Natural Plastics

Today many people consider only the synthetics as plastics. Actually the plastics industry started with animal horn and hoof, tortoiseshell, bone, ivory, gutta-percha, shellac, glue, and other compounds which necessitated the development of extruders, presses, molds, calenders, etc., which were used later for the synthetics as they arrived. Plastics were defined in the industry's first magazine *Plastics*, March 1926:

Plastic, as an adjective, is defined as the property of a substance by virtue of which it can be formed or molded into any desired shape, as opposed to non-plastic substances which must be cut or chiselled. Examples of Plastic materials, in that sense of the word, are clay, putty and ceramic-ware in general. Then there is the use of the same adjective as applied to works of art, and here the field is quite different, the word being usually applied to sculpture as opposed to two-dimensional representations such as drawing and painting.

Plastics, as a noun, seems to be somewhat of a stranger to most folks, but in these days of rapid development of new arts and sciences, such words soon become part of the language. So, to settle all arguments, we rise to remark that what we mean by "Plastics" is this: any material that by its nature or in its process of manufacture is at some stage, either through heat or by the presence of a solvent, sufficiently pliable and flowable, in other words,

plastic, so that it can be given its final shape by the operation of molding or pressing.

This includes a large variety of materials and hence enlarges the scope of the periodical which bears this name. The German language fortunately has an excellent word that describes all man-made raw materials and that is "Kunststoffe." We aim to have our magazine fill the same place among the English speaking world as the exemplary German publication that bears this name, but, for the present at least, are not including the very large field of rubber products, as this is ably covered by our contemporaries.

Thus the original generic term for the industry was *plastics* which has been shortened frequently to *plastic* in recent years. It is interesting to note that *plastics* is used in an 1862 mold patent to designate the material being molded. In 1929, the Molders Section of NEMA* conducted a contest to select a generic name for the industry and picked the name *synthoid,* but it was rejected by industry in favor of the

* National Electrical Manufacturers Association served as the original trade organization for plastics molders and the laminators.

word *plastics.* We shall include the early natural plastics in this chapter plus the natural plastics that came along in later years.

The earliest commercial plastics used in this country appears to be albuminoid, called keratin. Human hair, fingernails, horses' hoofs and animal horn are composed largely of keratin. Horses' and cows' hoofs, animal horn, and tortoiseshell were the principal raw material sources for these first plastics. Keratin fibers suitable for textile material and

Figure 1–1   Early use of plastics in the United States was in lantern windows. Glass replaced horn for lantern windows in 1740 after the opening of the first glass plant in the country. The window material is keratin, delaminated from natural horn. From this early start, developed the large comb business in Leominster, Massachusetts, where the early fabricators worked with horn, hoof, ivory, tortoiseshell, shellac, and Celluloid. (Courtesy, The Smithsonian Institution, Washington, D.C.)

made from chicken feathers were announced in the *Journal of Polymer Research* in 1946. It is interesting to note that Emil Fisher's much later work on nylon was initiated by a study to synthesize human hair. Adolph Spitteler called his 1897 casein plastics artificial horn and aimed at the markets for horn. Although horn was used as a fabricating material from the earliest days of man, there is no record to show when it was first used commercially in this country. Horners and hornsmiths fabricated these natural plastics. Horn windows for lanterns (often dialectically called "lanthorns") were replaced by glass soon after 1740, so that it is obvious that horn plastics were fabricated in the early part of the eighteenth century (fig. 1–1).

Horn is plasticized by boiling in water or soaking in an alkaline solution. While in this softened condition, it can be pressed into flat sheets, delaminated into thin sheets along the growth lines, molded into simple shapes, and it may be bonded to itself in a hot pressing operation. Its behavior is much like thermoplastic sheets. Horn can be colored and made to imitate the more decorative tortoiseshell. It was inlaid with metal, ivory, tortoise-shell, pearl fragments, and bonded to metal, wood, and other materials. All scrap can be reused. Hans Wanders, longtime secretary of the Plas-

tics Pioneers, entered the plastics industry in 1919 at Lawson Molding Company, where they molded ground horn and dried blood into fine plastics products.

Horn buttons were extensively made from cows' hoofs. The hoofs were ground, colored with a water solution, and compression molded like the thermoplastics. Hoofs contain a large amount of natural glue so the hoof molding powder bonded firmly when molded in hot dies and

Figure 1–2   Hand mold used in making horn buttons.

then chilled. Heated hand molds as shown in figure 1–2 were clamped on precut blanks and then held in vises while the dies cooled. In later years, hoof buttons were molded in 100-cavity molds. Rubber buttons (fig. 1–10) came along after the Goodyear patents in 1851.

Tortoiseshell buttons were made by softening the scales in boiling water, flattening them, and cleaning for the removal of impurities by scraping. Several of the scales were laminated to gain the desired thickness. Sometimes the scales were combined with horn or hoof blanks. Final shaping was done by molding in iron dies. Many of the variegated plastics today are made to imitate these early tortoiseshell products of the eighteenth century.

All of this developed into a large fabrication business with a high concentration on the very lucrative comb market. Noteworthy is the fact that hoof and blood glues were

Figure 1–3 Four combs of horn from Leominster, Massachusetts, which was known as "the comb city." The large ornate comb, made by John Evans Jones of Roxbury, Massachusetts, was presented to President Andrew Jackson. (Courtesy, The Smithsonian Institution, Washington, D.C.)

used for plywood bonding prior to the introduction of the phenolic adhesives.

The fabrication of "combs" was started in West Newbury, near Newburyport, Massachusetts, by Enoch Noyes. He is reported as the first comb maker in America, and his trade was making combs by hand from animal horns and tortoiseshells circa 1760. Enoch Noyes' primitive skills, gained from his family's experience in England,

were passed down through the Noyes people, and they eventually established substantial comb manufacturing businesses in Leominster, Newburyport, and West Newbury, Massachusetts, as well as Binghampton, New York. Leominster became the center of the comb business and was called "the comb city." Figure 1–3 illustrates the horn comb.

One of Enoch Noyes' first apprentices, Obediah Hills, moved to Leominster, Mass., from West Newbury in 1774.* The kitchen of Hill's house was his first comb shop. The work then, and for many years after, was all done by hand. For some years Obediah, Smith, and Silas Hills carried on this business, in a small way, in different places in town.

John Buzzell, who worked shell and ivory as well as horn, seems to have been the first to turn his attention to the question of tools for comb makers.

* This section is adapted with minor deletions from W. A. Emerson, *Leominster, Historical and Picturesque* (Lithotype Publishing Co., Gardner, Mass., 1888).

His labor-saving machinery caused many changes in the making of combs, improving the finish and increasing the facility of manufacturing.

Jabez Lowe, together with Charles and Thomas Hills, owned the first screw press, which for greater security against infringement, was built in a rude hut in Ashby woods before the patent was issued. The inventor was McPherson Smith, and the original patent papers are now in the possession of Thomas A. Hills. The patent was issued Jan. 28, 1818, and the papers were signed by J. Q. Adams,

Secretary of State. The press as built then is the same, in principle, as those in use in comb shops for the first century of plastics.

Jonas Colburn was the first man who used cotton cloth balls for polishing combs, and the first and only comb maker who made rolled-over combs, at one time doing considerable of that kind of work for other comb makers.

Ward M. Cotton invented the automatic machine, with cams, for cutting combs. He also made grails and other comb tools. The Buzzell quarnets and grails were also much used. The swing-jaw cutting machine was invented by Mr. Damon.

At first the horns used in this business were of small value. The manufacturers often brought the material, after cutting the horns into pieces for use, from Worcester on horseback. The value of the products was thus largely a matter of labor and so contributed most directly to the growth and prosperity of the town of Leominster.

In making horn combs, the horn was first cut into pieces with a com-

Figure 1–4   A collection of old-time combmaker's tools: (1) guillotine; (2) quarnet; (3) quiller; (4) shave; (5) bottoming saw; (6) gravers; (7) grile or grail; (8) topper; (9) wedge press. (Emerson, Leominster, Historical and Picturesque, *Lithotype Publishing Company, Gardner, Massachusetts, 1888.)*

mon handsaw, split open, and thrown into hot water to soften. It was then pressed flat by means of the old wedge press, which was one of the most conspicuous objects in all the early comb shops. This press was so arranged that when the pieces of the horn were put into place, they could be pressed flat by means of wedges driven in opposite directions. When this work was finished, the pieces were taken from the press and again softened by soaking in hot water. They were then ready for the next step, which was cutting of the teeth. This was done by means

of a small handsaw. A sharp knife was used to shave the comb; the smoothing and polishing was done by rubbing first with sand and water, and, after coloring, with chalk and vinegar or other preparations.

The bending was performed by using a number of small blocks of wood a foot long and three or four inches through. A small circular piece was cut from the side of each block of the exact size and shape desired, the comb put in its place, and a piece cut from the block put over the comb. The whole was kept in place by a stick

passed through staples in the sides of the blocks. In this way the comb was bent to the desired shape. When ready for market the combs were wrapped in the coarsest and poorest wrapping paper, in dozen packages, one of the number being placed on the outside as a sample.

The change which time has brought to this as to all other business, can in no way be more clearly illustrated than by recalling the fact that in the early days of comb making, it was not unusual for a man to make up what combs he could, pack them in saddle bags and start for the Boston market. In these days of rapid transit, of varied and perfect machinery, of large sales and profit, such a state of things seems hardly possible.

The old wedge press (fig. 1–4) was an important accession in the comb business previous to the introduction of the screw press. It was made from a piece of timber, mortised to receive the iron plates and strips of horn placed between them, and the wedges were driven down with a heavy beetle. The guillotine was used after the pieces were cut and straightened for

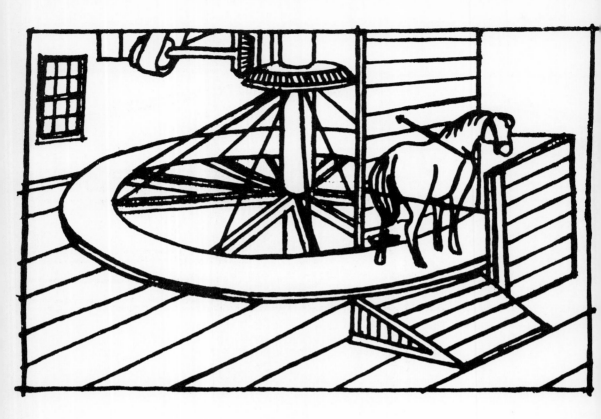

removing the wrinkles and flaws, the stock afterwards being still further reduced by the use of the crooked shave and the standing horse, the latter allowing a person to stand upright and being similar to what are used by leather finishers. The Topper or Pointer was in use when the ends of the comb teeth were cut off square, but went out of use when they were made by machinery. The Grile or Grail was used for rounding off and finishing the teeth, it being a tool of unusual application. The Quarnet was used principally on dressing combs to even and smooth the surface after shaving. Engraving tools of various shapes were used in all comb shops. The Quiller or Quilling Stock was used for cutting a bead on the plain stock before the comb was shaped, instead of the later process of embossing. The Bottoming Saw was used for rounding and sharpening the spaces between the teeth.

The old fashioned wheel horse power (fig. 1–5), as used in the early comb shops, is an interesting study in the light of modern improvements. They were located usually in the base-

Figure 1–5   This one horsepower engine was used in Rufus Kendall's Leominster, Massachusetts, comb shop early in the 19th century. The wheel, 20 ft in diameter, was supported by a center spindle from floor to ceiling and set at a slight angle. This was connected to gearing in the comb-fabrication shop. This stick over the horse was connected by a string to the floor above to expedite the horse, and a beam was used to drag on the shaft when the pace became too lively. (Emerson, Leominster, Historical and Picturesque, *Lithotype Publishing Company, Gardner, Massachusetts, 1888.)*

ment or first story, and the principal work was carried on in the story above. Mr. Rufus Kendall was the owner of the largest and most elaborately planned of these horse powers. It consisted of a round wheel or platform over 20 feet in diameter, supported by a large center spindle reaching from the floor to the ceiling, and set at a slight angle. This spindle was furnished with gearing which connected with the gearing on the main shaft. There was a stationary stall, one side of which was built from the floor, the other from the ceiling, with an in-clined walk, up which a horse was led into the stall, and a strap fastened across. In addition to the usual appliances, there was, in this instance, an interesting contrivance for increasing the speed when desired. It was a cherry stick about three and a half feet long, hinged to the side of the stall, and connected by a string leading to a small boy in the story above. When the power slackened, the boy worked the string, the horse struck a livelier gait and the machinery began to hum. To counteract the sudden increase of speed, a ponderous beam was sus-

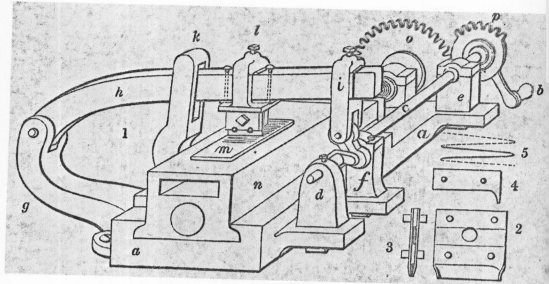

Figure 1–6 Machine invented by W. Rogers for making fine side combs from tortoiseshell: (aa) supporting plate; (b) winch attached to axle c; (de) upright pieces between which c and f, the crank of axle, are situated; (g) piece on which bar h hinges; (i) collar on bar h secured by screw and connected by means of a link with cranked part of axle; (k) loop which prevents bar h from swerving to either side; (l) cutter; (m) tortoiseshell out of which two parted combs are to be made.

pended from the ceiling, one end resting on a drum on the shafting overhead, so arranged as to drop and produce friction, thus regulating the speed when necessary.

Many machines as depicted in figure 1–6 simplified the process.

During the early part of the nineteenth century, rubber products were developed along with the advanced machinery that was later used for plastics.

George H. Richards of Washington, D.C., patented a process covering the use of a fluid rubber to waterproof fabrics. Edwin Chaffee of Roxbury, Massachusetts, who devised and patented the first calendering process (fig. 1–7) formed the Roxbury India Rubber Company in 1833. He manufactured coats, hose, life preservers, shoes, etc. The calendered vinyl business today is a direct offspring from that early work. Charles Goodyear's 1839 invention of a vulcanizing process

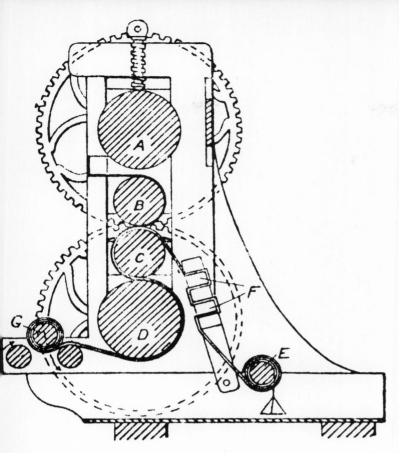

Figure 1–7   Edwin Chaffee devised and patented the first calender for Roxbury India Rubber Company in 1833 for the application of rubber to fabrics. This same process is used today for vinyl and other plastics.

for rubber developed that art of molding; the molds and the presses were subsequently used for shellac, gutta-percha, cold mold, and ultimately the synthetic plastics.

Rubber goods produced in 1850 were valued at $3,024,335, and this rose to $5,642,700 in 1860. At that time hard rubber was used for surgical and dental products, jewelry, buttons, dress ornaments, pencils, canes, etc.

Gutta-percha, a gum found in trees of the Malay peninsula and procured like rubber, was used concurrently for similar applications and for daguerreotype cases. Gutta-percha is similar to rubber except that it contains oxygen, which rubber does not have. Michael Faraday discovered the excellent electrical properties of gutta-percha even when it was permanently immersed in water. It became our first wire insulation and was the major insulation for ocean cable for many

years. The application of gutta-percha to the coating of telegraph wires was initiated by Mr. S. Armstrong of New York. The first machine was built in 1848, and the first wire so insulated was laid across the Hudson River at Ft. Lee in August 1849 for the Morse Telegraph Company.

The photo case shown in figure 1–8 had a basic position in our industry since the daguerreotype had just been developed and the de-

mand for pictures was very great. Samuel Peck is the first known plastics molder in the United States. He began working with the shellac plastics in 1852 and received his first patent (fig. 1–9) in 1854, patent number 11758. He entered into a partnership with Scovil Manufacturing Company in 1855 under the name of S. Peck and Company. Peck was bought out by Scovil in 1857, who continued the manufacture of daguerreotype cases and

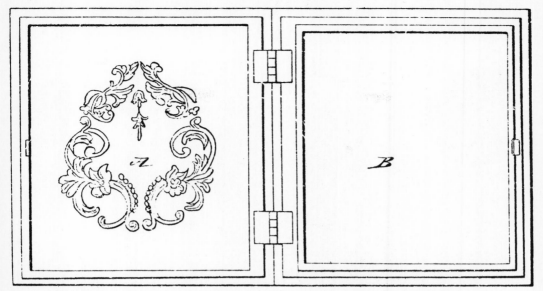

Figure 1–9   Drawings for patent of S. Peck, for "Making Photographic Picture Cases, No. 11,758. Patented Oct. 3, 1854. The composition of which the main body of the case is made, and to which my invention is applicable, is composed of gum shellac and woody fibers or other suitable fibrous material dyed to the color that may be required and ground with the shellac and between hot rollers so as to be converted into a mass which when heated becomes plastic so that it can be pressed into a mold or between dies and made to take the form that may be imparted to it by such dies."

other photographic specialties. Waterbury Button Company started molding buttons as a result of the nearby Scovil operations, and it continues today as a major producer of plastics buttons and other products for civilian and military use.

Little is known of Alfred P. Critchlow who is also credited as being an early American molder of plastics. He was a die sinker and horner who emigrated from England to Haydenville, Massachusetts.

He started by making horn buttons in 1843. Critchlow moved to Florence, Massachusetts, in 1845 and there developed the presses and dies to mold shellac and gutta-percha compounds. Critchlow's company, Florence Manufacturing Company, is now known as Pro-phy-lac-tic Brush Company (Standard Oil Company, Ohio) and is located where he started in Florence, Massachusetts.

These early plastics products (fig.

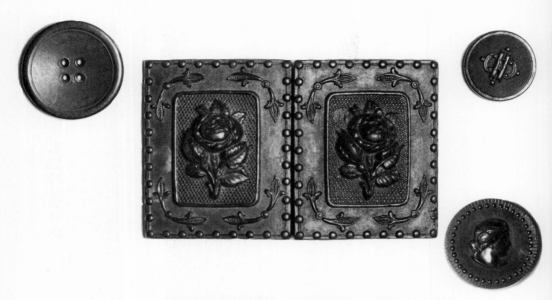

Figure 1–10   These early molded plastics products include shellac, rubber, and gutta-percha buttons and a shellac belt buckle.

1–10) included also jewelry and soap boxes, sewing kits, mirror backs (fig. 1–11), novelties, and dominos.

Shellac resins with various extenders continue to be used today for a variety of applications. They were used for compression, injection molded and laminated products in large volume (fig. 1–12) until after World War II, when the supply became limited. General Electric Company used a shellac binder for its Herkolite transformer insulation tubing long after the phenolics became available. Shellac was used as a binder for abrasive wheels and for dentures before the synthetics took over. It was the pioneer resin for many other applications that were converted to phenolics in later years. Shellac products had good wear resistance, high arc resistance, were nontracking, and their excellent fidelity in molding enabled them to serve the phonograph record market until

Figure 1–11   Mirror frame of molded shellac, circa 1868.

Figure 1–12   These complex shellac products were molded in 1927 by Western Electric Company. Shellac compounds provided good dimensional stability, mold-ability, and wear resistance. (From Western Electric News, *January 1927, page 9.)*

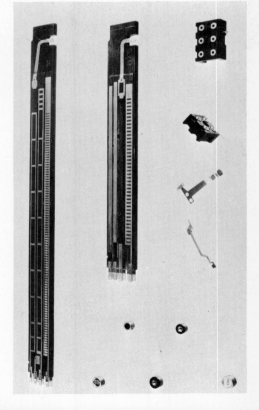

the better strength and the reduced noise level of the vinyl records pushed them out. During World War II, obsolete shellac phonograph records were ground up by RCA to salvage the material for new molded products. Some of the early molding compounds used sound like a "witches' brew." Patent #159,555, entitled "Compound for Picture Frames," calls for a material consisting of straw pulp mixed with one-half ounce of gum shellac, a quantity of alcohol to cut the gum, one-half ounce of glue, one-half pint of molasses, one-half pint of glycerine, and a quantity of ammoniacal solution of copper to make the pulp moist.

Another composition patent by Mark Tomlinson describes compounding and molding as it was done fifty years later: He describes the compound as

equivalents by weight of shellac, can-

Figure 1–13   Early screw press, used for compression molding in the Civil War era. Note the steam-heated platens.

nel coal and ivory black. The shellac and cannel coal are pulverized separately and the three ingredients are mixed and fed between steam heated rollers, one of which has higher velocity than the other, and are thereby ground into a pasty mass which while hot is cut to mold size and placed in a steam oven for 10–15 minutes and then molded and subsequently cooled for ejection.

An early press, circa 1860, is shown in figure 1–13.

A mold section of that era is shown in figure 1–14. The mold making in this era before Hyatt is quite fantastic when we consider their lack of machine tools and the exquisite design details achieved as illustrated by figure 1–15.

Plastics molding as developed by these pioneers was essential to the later works of Hyatt, Baekeland, Eichengrun, and the many other plastics pioneers who expanded the range of materials and markets into competition with every other mate-

Figure 1–14   Section of an old four-cavity mold, circa 1855. Note the mold force impression with KO marks. (Courtesy, Smithsonian Institution, Washington, D.C.)

Figure 1–15 Daguerreotype case with Landing of Columbus motif. Produced by A. S. Peck and Company from a painting by John Vanderlyn; die engraver, F. B. Smith and Hartman. No contemporary mold maker has done better work than this century-old molded part exhibits.

rial. Hyatt benefited immediately by the use of Peck's screw press with its steam-heated platens, Hancock's masticator, Bewley and Broman's extruder and Real's hydraulic press of 1816. Tremendous quantities of Celluloid sheets were made by slicing thin sheets from large molded blocks, using Hancock's slicer of 1840. The first submarine cable of 1850 was insulated by Bewley's extrusion process with gutta-percha.

It is interesting to note that the plastics molding industry was given its big start by the daguerreotype cases. Hyatt's Celluloid facilitated Edison's motion picture film. Eastman's cellulose acetate camera film popularized photography. The cameras of today, including the lenses, are made largely of plastics.

All of this early work simplified the mechanics for Hyatt and Baekeland, who brought in the synthetics. Noteworthy is the fact that it was a rubber molder, Richard W. Seabury, who envisioned Baekeland's varnish as a binder for molding compounds and who molded the first phenolic plastics.

The other natural plastics—cold mold, casein, lignin, soybean, zein, and furfural plastics—all followed this pioneering work much later, after synthetics had been developed.

Cold-Mold Plastics

Urgent demands for better electrical insulation and industrial molded parts by the rapidly growing electrical and automotive industries inspired the development of cold-mold plastics. The poor thermal endurance of shellac plastics limited its use in electrical and mechanical products. The more serious end-product problems had to be solved by the use of mica, glass, porcelain, rubber, ceramics, slate, wood and shellac-bonded paper laminates.

Emile Hemming, Sr., one of the important plastics pioneers, developed the art of cold molding in Europe around 1900 and, while an employee of Pathé Frères, created phonograph record materials and produced the first disc in France. In 1908, he came to America and pioneered cold-mold plastics in the field of electrical insulation. He established American Insulator Company in 1916 and founded Newark Die Company in 1917. He founded Plastics Products, Incorporated, in 1930, which he headed until his

death in 1964. Hemming's bituminous compounds had much better heat resistance and strength than did other contemporary materials. Prior to Hemming, plastics were molded in hot molds.

Cold-mold plastics were expanded in time with a variety of formulations to improve appearance, thermal endurance, dimensional control, moisture absorption, and strength. They had, and continue to have, many advantages. Molding is

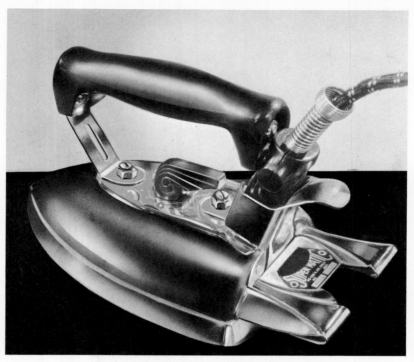

Figure 1–16 Early electric irons used cold-molded cord connector casings and control knobs.

achieved by a simple pressing and densifying in compression molds. Figures 1–16 through 1–20 show this work.

The cold-molded materials are generally divided into two classes: organic or nonrefractory and inorganic or refractory. The refractory cold-molded plastics were made from cement and asbestos with some clay. Another variety was formed also from slate and limestone. The organic or nonrefractory cold-mold materials use fillers of asbestos, diatomaceous earth, and binders of pitch, asphalt, and linseed oil; some synthetic resins are used in present-day compounds. Gilsonite, a glossy form of asphalt, is extensively used as a binder. One very large automotive application supplied by cold molders was the storage-battery box. Battery boxes used some combinations of the bitumens with fillers of asbestos, vegetable, and diatomaceous earth.

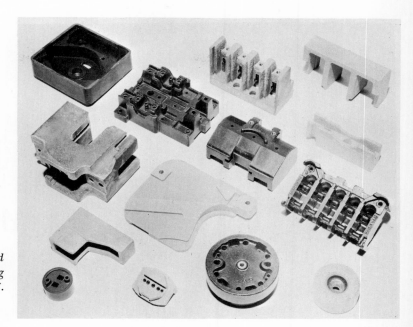

Figure 1–17   These cold-mold parts were molded during World War I.

Figure 1–18   These cold-mold parts were produced in 1970 and have a melamine additive to improve their arc-quenching properties. (Courtesy, Garfield Manufacturing Company, Garfield, New Jersey.)

Figure 1–19   Cold molding an inorganic plastics arc chute in an hydraulic press, circa 1935.

Figure 1–20   After the molding operation, these cold-mold plastics parts are air dried 3 hr, exposed to steam in a fog oven 24 hr, then immersed in hot water 24 hr, and subsequently baked 12 hr at temperatures up to 500°F. A wax additive was often added to minimize moisture absorption. Picture taken in 1940.

The bitumen plastics were melted and mixed with filler, and, while hot, the plastic mass was then put in the mold and pressed.

The organic cold-mold plastics were widely used general-purpose materials specified for wiring devices, switch components, and cookware handles. They hold up better under oven temperatures than the common thermosetting materials. A typical general-purpose formulation was made from raw linseed oil,

East India copal gum flour, manganese linoleate drier, stearin pitch, gilsonite, and linseed oil fatty acids. These raw materials were cooked (fig. 1–21) and agitated for many hours.

Curing this organic cold-molded material was a slow process. Small pieces were baked in an oven for approximately 25 hours, while large pieces were baked for a longer period. The baking process is complicated by the necessity for accurate

*Figure 1–21   Compounding the cold-mold bituminous plastics was dirty,
messy business. Photograph taken in 1940.*

control of the shrinkage and warpage. All of this was a messy, complex, and dirty business.

In the past, cold-molded materials served in many of the applications now filled by the thermosets.
Markets for cold-mold materials today are predicated on the following
outstanding properties: (1) high
temperature resistance, (2) excellent arc resistance, (3) low mold
investment, (4) low material cost,

and (5) new values gained by the
use of phenolic and melamine additives.

Glass-bonded mica (Mykroy, My-
calex, Havalex, Insanol) belongs
with this historical group of natural
plastics.

Percy B. Crosby, in 1919, was employed in England to find uses for
the mountains of scrap India mica
left after the best portions were
sold. He used a low-melt glass as a

Figure 1–22   Glass-bonded mica was the only good low-loss material available for all stable high-frequency apparatus in World War II. (Courtesy, Mykroy Ceramics Corp.)

binder and produced an amazing material just in time for the developing high-frequency, high-temperature needs of wireless transmitters and, later, broadcasting systems. Crosby worked with General Electric engineers at Schenectady in 1921 and 1922 perfecting manufacturing techniques. None of the presently available fine-organic, low-loss plastics materials were then available, and glass-bonded mica was the all-important, high-frequency insulating material for trans-

mitters and the better receivers as shown in figure 1–22. It was used in all Signal Corps apparatus in World War II. Teflon, the alkyds, styrenes, and other organics came along during the war and have filled many of these markets.

This pioneer ceramoplastics material continues to expand its service in the computer and other sophisticated electronics areas in which total dimensional stability (fig. 1–23) is needed at temperatures from absolute zero to 1300° F.

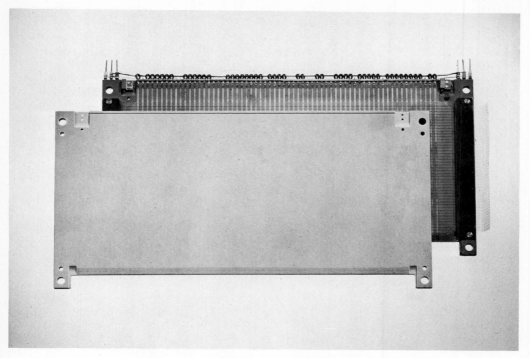

Figure 1–23   Absolute flatness and 25-year dimensional stability were achieved in 1970 by the use of glass-bonded mica for the substrate in the Bell System's revolutionary electronic switching system. No point in its 6″ × 12″ surface is allowed to vary from any other measured point by more than 0.006″. (Courtesy, Mykroy Ceramics Corp.)

Other contemporary applications make use of glass-bonded mica's total arc resistance, radiation stability, freedom from outgassing in vacuum, and low thermal expansion.

The cold-mold materials and the ceramoplastics bridge the gap between the organic plastics and the ceramics.

Casein Plastics

Casein plastics were developed by two Germans, W. Krische and Adolph Spitteler, in 1897. The first manufacturing of casein plastics in the United States started in 1919 by the Aladdinite Company. Others who followed were American Plastics Corporation and George Morrell Corporation. The business did not develop to a large size because of its poor moisture resistance and the greater abundance of other low-cost materials.

In the making of casein plastics, the enzyme rennet, which is present in the stomach of an unweaned calf, was used to precipitate the casein out of skim milk. Casein thus produced was washed and filtered, dried and ground. After mixing with appropriate plasticizers, this compound was pressed into cakes or extruded into profile shapes for fabrication by machining. Hardening was subsequently accomplished by immersion in a formaldehyde solution. It was used principally for

buttons and colorful dress ornaments.

Lignin Plastics

World War II with its shortages of materials introduced the use of lignin resins for plastics products. In this process, lignin material was separated from the waste sulphite liquors of paper mills and used to enrich the natural lignin content of wood cellulose. Thin sheets of lignin-enriched wood cellulose were molded by the process used for the laminated phenolics. Marathon Chemical Company of Rothschild, Wisconsin, introduced these products in 1941 as Lignin Laminated or Lignolite sheets.

Masonite Corporation of Chicago softened the lignin resin in chips of wood with 1,200-lb steam. Release of the steam pressure caused the chips to explode and produce a mass of small lignin-coated fibers which could be molded into sheets for construction and mechanical purposes.

Soybean Plastics

The U. S. Regional Soybean Industrial Products Laboratory at Urbana,

Illinois, studied the potential for soybean materials for plastics products in the middle thirties. When mixed with formaldehyde, soybean meal is moldable. Good products required the addition of phenolic resin also, and this was highly publicized by Ford Motor Company at one time, who used it for a few molded parts. It had high moisture absorption, poor dimensional stability, and was short lived as an industrial material.

Zein Plastics

Zein is a product found in corn, and, during the great depression years, many attempts were made to produce zein plastics by Quaker Oats, Corn Products Refining Company, and others. It was processed like the aforementioned soybean plastics and was also limited in application by its high moisture absorption. At one time it served as a flash drying resin for use in high-speed printing inks. A shellac substitute was also made of zein.

Furfural Resins

Furfuraldehyde resins were first produced in 1922 from oat hulls and patented by E. E. Novotny. The United States Department of Agriculture, Quaker Oats Company, and the Miner Laboratories worked together in the development of low-cost furfural materials from waste farm products. Furfuraldehyde plastics were extensively used during World War II since furfuraldehyde minimized the use of methyl alcohol for materials of the phenolic type.

2

The Celluloid Era

Almost every schoolboy has heard the fantasy about how the American plastics industry started: the genie in the drugstore bottle and the billiard ball contest. This widely publicized story has too often diverted attention from the facts and the truly great achievements of John Wesley Hyatt (fig. 2–1) and The Celluloid Company.

The burgeoning plastics industry in the days of the young Hyatt was seeking substitutes for shellac,

gutta-percha, ivory, hoof, India rubber, and horn. The record shows that the supply of rubber was seen then as very limited; gutta-percha (see chapter 1) had been substituted successfully for ivory in billiard balls a decade before Hyatt, but its supply also was limited. New plastics with abundantly available material supply were badly needed by the molders and fabricators of plastics. Many investigators were experimenting with cellulose nitrate, and camphor had been tried by others before Hyatt. The story is best summed up in the old adage: It isn't the thinker-upper who gets the payoff; it's the man who does something about it.

Hyatt worked diligently until he found the proper combination of cellulose nitrate and camphor to produce his easily fabricated, tough, and colorful thermoplastic—Celluloid (often called pyroxylin). He used all the contemporary plastics fabricating technology with imagination and dedication, achieving tremendous business success and greatly accelerating the plastics evolution. The machine and process developments that were refined and expanded for the processing of Celluloid are considered by many to be as important as the Celluloid itself.

Recognition of the solvent action of camphor on cellulose nitrate plus the machinery inventions are significant factors which gave The Cel-

Figure 2–1   John Wesley Hyatt (1837–1920), inventor of Celluloid and many of the plastics processing methods.

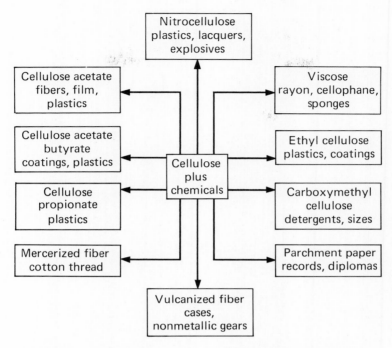

*Figure 2–2   The cellulosic plastics. (Courtesy, E. I. du Pont de
Nemours and Company, Incorporated.)*

luloid Company and John Wesley Hyatt worldwide recognition. Theirs was the first good semisynthetic plastics material and the start of the cellulosic plastics (fig. 2–2).

A friend of John Wesley Hyatt, Peter Kinnear, who lived in Albany, assisted in the formation of the Hyatt Billiard Ball Company with headquarters at Kinnear's home. The name was soon changed to the Albany Billiard Ball Company, and the firm concentrated on

making composition billiard balls with little or no cellulose nitrate being used. Their first profitable item using Celluloid was an item of costume jewelry, which they marketed as South American jewelry. The Hyatt backers, General Marshall Lefferts, Joseph Larocque, and Tracy R. Edson, influenced Hyatt to move to Newark, N. J., and form The Celluloid Manufacturing Company. During the next forty years, Hyatt, who did not participate

The Chas. Burroughs Co 156

Figure 2–3 Slugs of cellulose nitrate were molded in this press by The Celluloid Company. (Courtesy, The Charles Burroughs Company, Newark, New Jersey.)

in the business management, developed products and methods; 250 patents were issued to John Wesley Hyatt in many fields, the Hyatt Roller Bearing being one of the most famous.

The Burroughs Company, established in 1869, prints this message in its 1926 catalog:

Sixty years ago, Mr. John W. Hyatt made the material now known univer-sally as Celluloid. At that early period, he sought the services of a practical engineer to collaborate with him in the design and construction of proper machinery for manufacturing his ma-terial and the various articles that might be made therefrom. The engi-neer engaged was Mr. Charles Bur-roughs, the founder of the Burroughs Company. Since then, this company under his direction and that of his successor, Mr. Charles F. Burroughs, has been constantly engaged in the de-sign and manufacturing of machinery

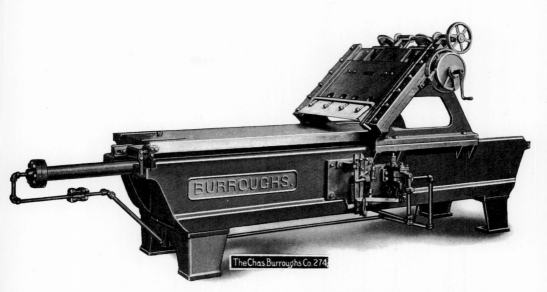

Figure 2–4   Celluloid slugs were sliced into thin sheets in this Burroughs Hydraulic Planer. (Courtesy, The Charles Burroughs Company, Newark, New Jersey.)

and tools for working, not only Celluloid, but practically all plastics known to the trade today.

Hyatt's vision of improved manufacturing processes, often reduced to practice by Charles Burroughs, became the production backbone of the plastics industry. These machines permitted immediate and widespread production of phenolic plastics after Baekeland. While Celluloid could not be injection molded as defined today, it was injected by a piston extruder into molds, or compression molded, into slugs (fig. 2–3) that could be sliced or planed into thin sheets for fabrication. Burroughs' compression sheet molding press and the hydraulic planer (fig. 2–4) were widely used for the making of thin sheets. Sheet extrusion was developed many years later.

Figure 2–5   Celluloid baby rattles show the intricate detail that could be molded into decorative and functional articles. These doll-like figures, precursors of the famed Kewpie doll, were among the first plastic products (circa 1890) to be made by the Hyatt blow-molding technique.

Hyatt himself describes what probably was his start of injection molding.

The next step was to make a small stuffing machine consisting of a cylinder, 4 inch bore, about a foot long, terminating in a tube, ¾ inch bore, 10 inches long, immersed in an open oil jacket with a gas burner and thermometer. A cap nut forced a plunger upon cakes of incipient Celluloid which were heated at the outlet end of the cylinder, passing through the heated tube into molds and also through nozzles forming rods, tubes, etc.

Hyatt used Bewley's 1845 extruder design illustrated in figure 8–1. Noteworthy is the fact that we followed this piston extrusion procedure plus the torpedo in our injection machines for thirty years before going to reciprocating screw injection.

Hyatt also conceived a practical

Figure 2–6   Hyatt conceived Celluloid blownware by expanding tubing in a blowing press. (Courtesy, The Charles Burroughs Company, Newark, New Jersey.)

blow molding process wherein heated Celluloid tubes or sheets (fig. 2–5) were expanded to fit mold contour. A generation of toys, baby rattles, vanity products, packages, and decorative items were blow formed by this process. The popular Burroughs Blowing Press shown in figure 2–6 is a far cry from today's hot-melt blow molders, but it served to develop the art and to provide basic ideas and inspiration for Ferngren's, Kopitke's, Bailey's

and Moreland's later work with the synthetic plastics (chap. 8).

The first chapter tells how Leominster, Massachusetts, became the center for the fabrication of Celluloid products and came to full fruition in the period from 1900 to 1920. Automatic equipment, developed in Leominster, was sold and imitated throughout the world. Leominster became also a center for making plastics presses, molds, and fixtures. It is probable that more

Figure 2–7   A shellac mirror back from the days before Hyatt is shown on the left. The Celluloid-covered mirror, comb, and blown hair container were very popular items in the Celluloid era.

people were employed in plastics in Leominster during that period than are employed today throughout the industry since there was less automation at that time. More hands were needed. Many of these workers were real artists as is evidenced by the beautiful jeweled Spanish combs and art novelties which they created from Celluloid.

The gay nineties and the teens of the twentieth century were great days for Celluloid. At the peak of the Celluloid Era, production was recorded at 40,000 tons per year. In his 1964 book, when the British celebrated *The First Century of Plastics,* Mr. M. Kaufman reports,

more than forty quite different types of applications were produced, such as linens, knife handles, dentures, fancy goods, printer's blocks, and tooth brushes.

The "linen" applications included celluloid collars, cuffs, and dickeys. Combs [fig. 2–7] were very high-volume items; they moved

Figure 2–8   This storm front for the buggy of the gay nineties was the predecessor of the early auto window for wind and rain, which was detachable and made of Celluloid.

Figure 2–9   Early motorists' dust goggles like this pair had Celluloid eye windows.

from a useful necessity to a high-style item. Corset stays, shoe-heel coverings, printed signs, side curtains [figs. 2–8 through 2–10] spectacle frames, and piano key coverings were all big business. Celluloid replaced ivory and was used for the natural keys on pianos until 1950. Piano sharps were thermo-formed Celluloid sheets on wood cores. Some spectacle frames are still made today of Celluloid.

Mr. Kaufman also quoted Dr. Fassoti of Mazzucchelli as saying,

"Women, round the turn of the century, carried a couple pounds of Celluloid about their persons in the shape of many combs." A resulting major collapse of the Celluloid business came in 1923 when Irene Castle's bobbed hair became the thing. The Celluloid dental plates that replaced shellac were introduced by Hyatt and were used widely until replaced by acrylic and other resins many years later.

The Reverend Hannibal Goodwin, rector of an Episcopal church

Figure 2–10   Model T Ford of the 1920s. The rain curtains were of cloth coated with cellulose nitrate, and the windows were of Celluloid.

in Newark, was desirous of making his own stereoptican slides for church lectures and conceived the idea of substituting Celluloid film for the glass plates. His patents, filed in 1887, were acquired by Ansco and later sold to Eastman Kodak Company. The amateur photographic business (fig. 2–11) was accelerated greatly by the displacement of glass plates with roll cellulose nitrate film by Eastman Kodak Company. The plastics mold-ing business, which started with daguerreotype cases (fig. 2–12), was expanded by the film business and today even the lenses of many cameras as well as the cases are made of plastics. Plastics and photography have grown up together.

John Henry Stevens, in charge of Celluloid research, first suggested amyl acetate as a cellulose nitrate solvent and thus fathered the nitrate lacquer business. As coinventor in 1892 with The Celluloid

(a).

(b).

"You press the button
We do the rest"

Jack: Do you think baby will be quiet long enough to take her picture, mamma?
Mamma: The Kodak will catch her whether she moves or not ; it is as " quick
as a wink."

Send to the Eastman Company, Rochester, N. Y., for a
copy of " Do I want a Camera," (illustrated) free by mail.

(c).

Figure 2–11   Celluloid film made possible motion pictures and
low-cost photography with the perfection of a process for making
the tough, clear film in continuous lengths by John Stevens (1853–
1932), whose portrait is shown here (a). Film cameras were advertised as early as 1890, and motion picture films were produced by
1892. The Celluloid motion pictures frame given here (b) shows
an unknown ballet dancer from a movie produced by the Edison
Company. Also shown (c) is an early Kodak advertisement (about
1890). Plastics sheeting replaced glass plates, thus making possible
the portable camera.

Figure 2–12   Shown here is the transition from the molded shellac plastics of the days before Hyatt to a Celluloid-covered wood case with pearl inlay.

Company president, Marshall C. Lefferts, of the machine for casting continuous cellulose nitrate film on a rotating drum, he deserves high recognition in the archives of plastics machinery invention. The film casting machine patents were sold to Eastman Kodak Company. Edison's success with the motion picture machine was predicated on a tough clear film, and cellulose nitrate did that job until replacement started with the slow-burning acetate material.

Eastman Kodak Company first produced triacetate film in 1908; they made acetone-soluble products in 1909–11 and in 1917–18 made cellulose acetate for wing-covering dope. In 1925, the Kodak Research Laboratory concentrated on work leading up to the replacement of cellulose nitrate by cellulose acetate in all film types.

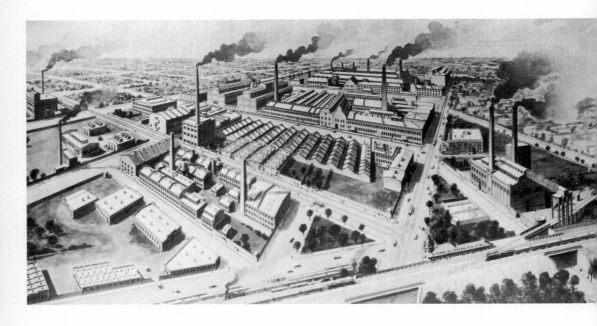

Easy workability, excellent colors, water resistance, toughness, printing ease, clarity, cementability are the characteristics that contributed to the greatness of the Celluloid Era (fig. 2–13). Celluloid's lightninglike burning property is the one defect that prevented its continued use and stimulated the development of nonburning cellulose acetate and the phenolic plastics. The many disastrous Celluloid fires made the underwriters very sensitive to plastics that were not self-extinguishing.

The Celluloid Era was worldwide, and any report on Hyatt's impact must mention the many companies that were spawned by his success. Celluloid's predecessor company in England, Parkesine Company, made little progress until 1877 when it turned to Hyatt's formulations. Parkesine's successor, the British Xylonite Company, prospered and became big business.

igure 2–13   The Celluloid Company.

Plants were soon established in France, Switzerland, Germany, and Japan. In the United States, Hyatt's success in Newark initiated the Merchant's Manufacturing Company, in 1881, which by 1915 became the largest Celluloid maker. It was purchased by E. I. du Pont de Nemours and Company (Du Pont) when they made their entry into plastics. The Fiberloid Company, which started in 1894 to make collars and cuffs, was purchased by Monsanto for its start in plastics. Other early entries into the plastics industry via Celluloid were Celanese Plastics Company, Foster Grant Company, Eastman Kodak Company, Hercules Powder Company, and Nixon Nitration Works.

Many conflicting stories have been presented in connection with the development of cellulose nitrate plastics and the following study was prepared by Dr. P. W. Bishop of the Smithsonian Institution to provide

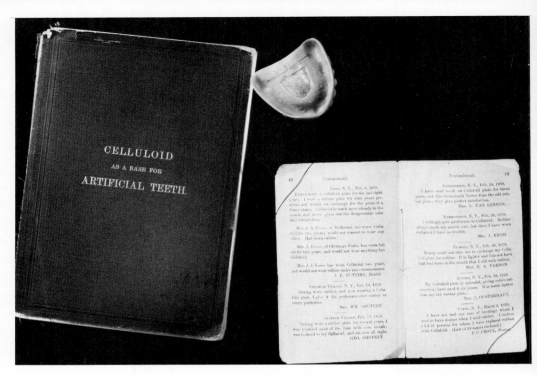

Figure 2–14   Celluloid dentures gave impetus to the early growth of plastics when dentists sought a replacement for vulcanized rubber, wood, and even metal dentures. Testimonials from patients and dentists were gathered into a book printed about 1880. Shown here are a celluloid denture (circa 1875) and excerpts from a booklet issued January 1881 by The Celluloid Manufacturing Company entitled Artificial Teeth Testimonials from Dentists.

a true perspective of this great inventor's work.

The offer of a premium of $10,000 to the inventor of a substitute for ivory billiard balls attracted the attention of John Wesley Hyatt. His experiments, which led to his first patent (USP 76765) dated April 14, 1868, did not result in invention of Celluloid.*

* The plastics industry unwisely selected 1868 as the start of its Centennial—J.H.D.
He combined two materials, already well known. A material which may

very well claim to be one of the earliest plastics and which had been used commercially since the early 1850's formed the body of his billiard ball. The combination of fibrous pulp with gum shellac, for which a patent was issued to Samuel Peck in 1856, was used to make daguerreotype cases on a large scale. These cases [fig. 1–8], of which thousands survive in the hands of daguerreotype collectors, testify to the response of this material to moulding under pressure. Hyatt's claim in this connection was that he had im-

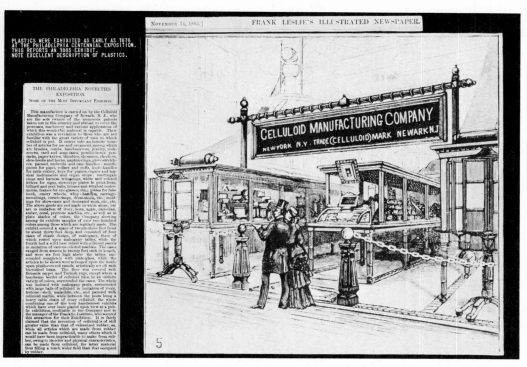

Figure 2–15   Celluloid on Exhibit began at the Philadelphia Centennial Exposition in 1876. The Celluloid exhibit for a Philadelphia Exposition in 1885 was reported by Frank Leslie's Illustrated Newspaper *as "one of the very handsomest exhibits which have ever been placed upon view at a public exhibition."*

proved the process by pulverizing the gum shellac before mixing it with the pulp.

His next step was to cover the moulded ball with collodion, a material which had been in use since 1855. This thick colorless liquid is obtained by dissolving in a mixture of ether and alcohol, a form of cellulose known as pyroxyline. The latter was made by mixing purified cellulose in a mixture of sulphuric and nitric acid. The evaporation resulting from exposure of collodion to the air leads to the formation of a film of pyroxyline. When this was made flexible by the addition of some castor oil, it became effective as a cover for wounds, a 19th century predecessor of the modern flexible bandage.

As E. C. Worden, a recognized authority on the nitrocellulose industry, and various writers in *Plastics* have pointed out, the invention of nitrocellulose by Schönbein in the 1840s stimulated many experiments to adapt cellulose nitrate to broader applications than explosives. One of the pro-

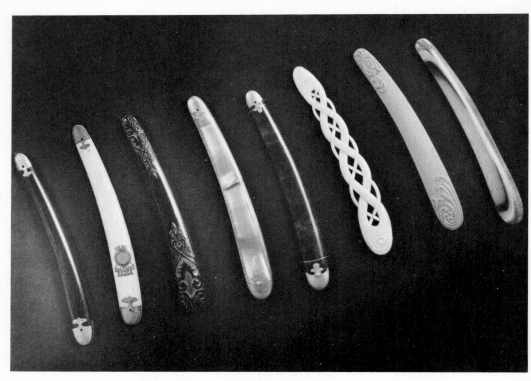

Figure 2–16   Straight razor handles of Celluloid were made to simulate onyx, carved ivory, ebony, bone, marble, and silver-inlaid pearl or shell in the era from 1880 to 1910.

ducts which emerged was collodion, used first in medicine and, later, in photography. Many attempts were made, in the United States and elsewhere, to make the collodion film flexible for use as a waterproof coating and for moulding into dental plates; and it is difficult to believe that Hyatt was unaware of some of this work.

Hyatt does not record that his new billiard ball received the premium. From the fact that as late as 1884 the Scovill Manufacturing Company of Waterbury, Connecticut, obtained a large contract for pool balls made

simply of the Samuel Peck composition, it is reasonable to assume that Hyatt's own Albany Billiard Ball Company was successful only when the collodion covering was omitted; for Hyatt himself reported later that

we had a letter from a billiard saloon proprietor in Colorado mentioning this fact [that occasionally the violent contact of the balls would produce a mild explosion like a percussion gun cap] and saying that he did not care so much about it, but that instantly every man in the room pulled a gun.

Hyatt's experience with collodion

Figure 2–17 Early molding of plastics for elaborately carved handles of comb-and-hand-mirror set made of Celluloid in 1880 demonstrates the infinite variety of shape and contour that becomes possible with plastics.

introduced him to the problems of its manufacture if, that is, its characteristics were to be made use of in commerce. He was joined by Isaiah in 1869 and, according to his own account, serious work then began to solve the problems inherent in the material.

Coincident with his account of his initial experiments, apparently designed primarily to overcome the risk of explosion, Hyatt remarks

. . . finding it stated in some patents to which I was referred, *that a little camphor added to the liquid solvent*

was beneficial, we conceived the idea that it might be possible to mechanically mix solvents with the pulp and coloring matter while wet, then absorb the moisture by blotting papers under pressure and finally submit the mass to heat and pressure [emphasis supplied].

This remark can hardly be used to affirm that Hyatt had looked at the patents which were the subject of the later litigation; but, at least, it permits one to question the assertion by Professor Charles F. Chandler that Hyatt "was entirely ignorant of the

Figure 2–18 Smokers' accessories were molded of Celluloid plastic in the early 1900s. The cigarette box at the top was presented by John Wesley Hyatt about 1915 to George Wald, a pioneer plastics chemist.

various efforts which had been made by Parkes [sic], Spill and others to utilize soluble nitrocellulose or pyroxyline for the manufacture of plastics materials." Again, it seems reasonable to assume that Hyatt, as a prudent business man did, in fact, follow up the lead; and that he was not ignorant of the efforts of others in the field. Who were these others?

Alexander Parkes (1813–1890) was a metallurgist born in Birmingham, England and for a long period associated with Elkingtons, the originators of electro-plating. He was a prolific innovator with 66 patents to his name, most of them relating to metallurgy and, especially, to electro-metallurgy. His excursion into the non-metallic world followed his introduction, in 1841, of a process for waterproofing fabrics. This, in effect, was a development of the method of vulcanizing rubber patented by Goodyear in the United States and Hancock in England. Their process treated natural rubber with sulphur at high temperatures; Parkes introduced a means of rubberizing cloth with a cold solution, his patent for which was subsequently

Figure 2–19   Reading and writing were aided by pen barrels and eyeglass frames molded of Celluloid plastic. The pince-nez eyeglasses date from the turn of the century and the multicolored fountain pen from about 1932.

taken over by the famous waterproofing firm of Macintosh. Other important patents followed and were taken up by the rubber industry during the 1840s; and it is probable that the interest which this non-chemist derived from his experiments with rubber encouraged Parkes later to pursue an idea that a "new material . . . was much required" by industry. It seems to be agreed that the immediate stimulus to Parkes' experiments was the behavior of collodion when exposed to the air. By 1862, Parkes was able to exhibit objects made from "Parkesine"

at the Great Exhibition of 1862 in London.

Parkes' patents covering the new material were granted between 1855 and 1865. The business started by Parkes in 1866 as the Parkesine Company operated haltingly till 1868, his failure to succeed commercially being attributed, in part, to an overconcern for low costs, leading to the use of poor quality raw materials. Since Parkes lived more by developing processes than by operating them on a commercial basis, it is perhaps not surprising that this venture into production was unsuccessful, and led to an association with an established business man, Spill.

Daniel Spill (1832–1884), waterproofer, who was awarded a medal for his products in the Great Exhibition (London) in 1862, became interested in Parkes' product as a waterproofing agent and worked on it at his factory before becoming more directly associated with Parkes, in 1865, as the works manager of the Parkesine Company. When the latter closed down and Parkes abandoned his interest in his "new material," Spill continued to produce what he called "Xylonite" and "Ivoride." He took out two British patents in 1867 and 1868 (among others) and in the following years received corresponding United States Patents (97,454, in 1869 and 101,175, in 1870).

The essential problem which confronted these pioneers was the choice of a solvent which would permit conversion of pyroxyline into a plastic mass capable of being moulded under pressure and heat; and the essence of the argument which occupied the Federal Courts for a number of years was whether Hyatt, Spill or Parkes discov-

ered that camphor was the essential ingredient; and whether Hyatt knew that his English predecessors were, at least, not unfamiliar with camphor's attributes in this connection.

Hyatt's patent of immediate interest to this story was granted in 1870, and the Albany Dental Plate Company, which made the first commercial use of Hyatt's patents, was established in the same year.

Spill, apparently, took no steps to establish an operation in the United States until 1881, after the U.S. Federal Court had apparently established the validity of his patents. Meanwhile, in 1872/3, when Hyatt's factory was moved from Albany to New York, the name *Celluloid Manufacturing Company* first appears.

An interesting aspect of the first stage of the litigation in which Spill sought to prove that Celluloid had infringed his patents was that he had retained as an expert Professor Seeley. The latter's views are quoted with approval by the Judge in his decision of May 25, 1880. The same Seeley is referred to by Hyatt on the occasion of his acceptance of the Perkin Medal as one who

had made collodion for the govern-
ment during the Civil War. He came
to our place in Albany, N.Y. [on be-
half of the American Hard Rubber
Co., whom the Hyatts were hoping to
interest in their product] and we con-
ducted the whole process for his in-
spection, very successfully.

An early example, indeed, of the involvement of an academic chemist in
commercial affairs.

On May 25, 1880, the Federal Circuit Court of New York gave judgment in favor of Spill to the effect that
Celluloid had infringed his patents
USP 97,454 and 101,175. The defense,
said the opinion of Blatchford, C.J.,
had failed to show want of novelty; or
to establish that Spill had, in fact,
obtained from Parkes his knowledge
that alcohol and camphor united were
a solvent of xyloidine or pyroxyline.
As to Spill's use of bleaches to treat
pyroxyline, the court could find nothing in Celluloid's defense to counteract the view of Spill's expert, Professor Seeley, that Spill had, by his
invention, applied ordinary bleaching
agents with "novel and useful results
. . . by such simple and unlooked for

methods"; nor again, had the defense
shown that Spill had obtained this information from Parkes.

Apparently as a consequence of this
success the Spill patents or a license to
use them was acquired in 1880/1 by
an American group which formed the
Xylonite Company with a factory in
Adams, Mass. The record is unclear as
to who handled the subsequent litigation, though it was conducted in the
name of Spill.

The interlocutory decree which followed the 1880 decision awarded a re-
covery of profits and damages "to be
ascertained by a master." After three
and a half years of contemplation, the
master reported on February 25, 1884
that "not having been furnished with
the necessary data" he could not, without further proof, report any profit
derived by Celluloid from the infringement of USP 97,454; that he was
in the same position with respect to
the infringement of USP 101,175; and
finally no evidence had been sub-
mitted with respect to the damages
claimed. Spill's attorneys countered
with a series of exceptions to the
master's report, claiming that profits

amounted to $780,973.91, with interest from June 12, 1880.

Celluloid argued that the whole case should be reconsidered, having regard to further evidence "since introduced into the case" and in view of the Supreme Court's decision in *Pennsylvania Railroad Co.* v. *Locomotive Engine Safety Truck Co. 110. US 490.* in which it was established that the application of an old process was not patentable.

Whether or not Justice Blatchford was unduly influenced, as Xylonite's president Emil Kipper appears to have believed, he proceeded to review his previous decision and to come down handsomely in favor of Celluloid. The judgment of August 21, 1884 reviewed in great detail the Parkes patents and came to the conclusion "that due weight was not given to these considerations, in connection with the state of the art . . . and that [the interlocutory decree] ought not to have been made" in respect of USP 97,454; moreover, being bound by the Supreme Court's decision in the Pennsylvania case [cited above] he ruled that "the application of an old process . . . with no change

in the manner of application and no result substantially distinct in its nature, will not sustain a patent, even if the new form of result has not before been contemplated."

In effect the Judge's decision meant that Spill's claims lacked novelty because the ideas had already been covered by Parkes in his English patents; and, it seems, that Alexander Parkes, himself, approved of this point of view.

This decision did not end the litigation. American Xylonite continued in production and in 1886, Celluloid sued them for infringement of their USP 156,353 of 1874. An injunction was granted on March 5, 1886 in spite of Xylonite's claim, based presumably on Blatchford's decision of 1884, that this patent also was defective for want of novelty. Justice Shipman added a comment to his decision which, in effect, ignored Xylonite's own contribution to the art. He said:

To the process set forth in the patent and the knowledge and skill which grew out of an acquaintance with it, is due the present commercial success of xylonite or celluloid, as an article which can be devoted to a very great variety of uses.

In a petition for rehearing, decided on July 27, 1886, the fine distinction between Hyatt's and Xylonite's methods was further clarified. In rejecting the petition, Shipman said:

The defendant (Xylonite) cannot escape the charge of infringement by the circumstances that it first abnormally *dries the wet pulp, whereas the plaintiff first substantially or* comparatively *dries it and then after the pulp and camphor have been mixed, expels all the remaining moisture. By such an alteration of an important and valuable process infringement is not avoided.*

Again, he emphasized that "where twenty years of inventive skill of no ordinary character and of different persons had been most earnest and persevering in the effort to produce such improvements, the patentability of the improvement when produced cannot be doubted." Hyatt's improvement, according to the Judge, was the first to make an article which should be both attractive and useful. American Xylonite continued in operation

at its plant at Adams, Massachusetts, till January 1891, when Celluloid took over the business. At this time Xylonite employed 500 men and it is a fair assumption that their techniques must also have been the result of earnest and persevering effort! . . . Celluloid did not have plain sailing in the patent field as a result of its success against the 1886 petition; for as late as 1890, their claim against the Cellonite Manufacturing Company for infringement of USP 156,352 was unsuccessful.

It is curious to note that while the initial impetus to Hyatt was the search for a substitute for ivory, his patent (USP 89582, May 4, 1869) was for a formula for combining ivory dust with collodion in the proportions of 3 : 1, the mixture being shaped under pressure of up to 10,000 pounds per square inch. Here we have the first hint of a possible explanation of the success of the non-chemist in this chemical field; and of the involvement in chemistry of an engineer whose career culminated in the invention of the roller bearing. The makers of the so-called "Union" case for daguerreotype had, in the 1850's, considerable difficulty with their presses and dies; and, undoubtedly, Hyatt had the same difficulties till he and his brother could evolve adequate machines to mould his Celluloid into useful objects.

Their next patent (USP 91341, June 15, 1869) emphasizes this; for now they propose to make solid collodion by dissolving the pyroxyline in the press by applying the ether or other solvent in the mould itself, using pressures of between 10,000 and 40,000 psi to force the solvent into contact with every particle of pyroxyline.

The next stage and the essential prelude to the manufacture of celluloid as opposed to solid collodion is described in the patent (USP 105,338) dated July 12, 1870, which undoubtedly provided the basis for the first commercial enterprise of the Hyatts. In short, the Hyatts added to their ground pyroxyline pulp, while still wet, finely pulverized gum camphor in the proportions of 1 part camphor to 2 of pyroxyline. The camphor, according to their claim, was capable of being vaporized or liquified and converted into a solvent by the applications of heat and applying heavy pressure while heated. This became the Hyatts' basic patent. It was reissued on June 23, 1874 (and numbered 5926) and modified slightly in USP 156,353 (Oct. 27, 1874) in which the quantity of camphor was reduced from 50% to between 25 and 40% of the quantity of pyroxyline with, also, a reduction in the degree of heat.

Enough has been said to show that, without denying Hyatt the credit due to him for his successful solution of fundamental problems, and for his establishment of a great industry in America, he, like all inventors, must be fitted into the complicated pattern of evolution which is the story of technology. It is probably true to say that a seventeenth or even eighteenth century chemist could build himself a house in a tree and, in isolated contemplation, discover a new element; but in the ferment of nineteenth century ideas, no man worked in isolation. Crude equipment had given way to accelerating sophistication in design and, especially, to design for special purposes. The Hyatts' success as amateurs in chemistry came, undoubtedly, from their expertise in the mechanical world. The name *Celluloid,* the inspiration of Isaiah in 1872, survives to commemorate the Hyatts' work, even if the product has taken a back seat to an ever-expanding list of new materials.

The Albany Billiard Ball Company founded by the Hyatts continues today as the leading manufacturer in this field. The first successful production balls were made of bone filler, camphor, cellulose nitrate, and coloring. These materials were made into sectional cold preforms, the main body, colored ring, and the number. Such pieces were then fused together in a molding process wherein one section was molded on top of the next, step at a time, until the complete ball was produced. Finishing operations included diamond turning, centerless grinding, and wax polishing. In subsequent years, phenolic casting resins as described in chapter 3 were used. Noteworthy is the fact that Albany Billiard Ball Company did double-shot and quadruple-shot

molding long before such patents were issued to molders of thermosets and thermoplastics.

Early Cellulose Chronology

The following chronology is taken from *Cellulose, the Chemical that Grows* by Williams Haynes, copyright 1953 by Williams Haynes, copyright 1953 by American Association for the Advancement of Science. Reprinted by permission of Doubleday & Company, Inc.

Before 3500 B.C.   Egyptians made papyrus from pith of aquatic reeds.

Before 2600 B.C.   Natural silk fibers spun and woven by Chinese.

Before 100 B.C.   Paper invented by Chinese.

751 A.D.   Chinese art of papermaking revealed to Arabs at Samarkand.

1664   Dr. Robert Hooke suggested

"artificial silk" in his *Micrographia*.

1690 William Rittenhouse, German papermaker, and William Bradford, English printer, partners in first North American paper mill, near Philadelphia.

1710 Réaumur, French scientist, pointed out possibility of artificial fibers from gums and resins.

1740 Gloves and stockings knitted of spider web, Bon (France).

1770 Dubet described method of drawing out filaments from the gum of dead silkworms (France).

1798 Papermaking machine invented, Louis Robert (France).

1833 Xyloidine (nitrocellulose) prepared by reaction of nitric acid with starch, linen, and sawdust, Braconnet (France).

1838 Pelouze (France) treated paper with nitric acid to get similar product.

1839 Payen (France) isolated cellulose in wood.

1840 Mechanical wood pulp (ground wood) used for papermaking, Keller (Saxony).

1842 Filament spinning machine suggested by Louis Schwabe (England).

1844 Mercerizing process by action of caustic soda on cotton textile fibers discovered, John Mercer (England).

1846 Cellulose trinitrate made commercially from cotton by treatment with mixed nitric and sulfuric acids, Schönbein (Switzerland).

1847 Ether-alcohol solvent for nitrocellulose, collodion, Louis Ménard and Florès Domonte (France). Böttger and Otto confirmed Schönbein's preparation of nitrocellulose from cellulose and nitric acid. Military-chemical commissions named in France, England, Russia, Germany, and Austria to study nitrocellulose as explosive.

1848 Collodion applied to medical uses, J. Parker Maynard, Boston, Mass. Explosions in Hall Bros. nitrocellulose plant, Haversham, England, and at Brouchet, France.

1850 Collodion emulsion base for light-sensitive chemicals on a photographic plate invented independently, F. Scott Archer (England) and Gustave Le Gray (France). "Flexible collodion" made by adding Venice turpentine as a softening agent, C. S. Rand, Philadelphia. Shellac-wood fiber plastic composition, Halverson (U.S.). British and French commissions investigating guncotton disbanded, and Russia forbade manufacture or import of nitrocellulose.

1852 First pyroxylin lacquer exhibited at London Exposition, Alexander Parkes. Von Lenk perfected safer process for nitrating cellulose and Austrian Army adopted guncotton as propellant explosive.

1855 Silklike fiber of nitrocellulose prepared from mulberry leaves and rubber, George Audemars (Switzerland). First patents for cellulose plastic, Parkes (England).

1856 Spirits of camphor used as nitrocellulose solvent, Parkes. First patent for pyroxylin-coated fabric, leather substitute, Green (England).

1857 Schmetze (Germany) separated cellulose and lignin in wood and named both. Solubility of cellulose in ammoniacal solution of copper discovered by Schweitzer (Germany); basis of cuprammonium rayon process.

1858 Ground wood-pulp paper first made in U.S., Burgess and Watt.

1862 Parkesine, nitrocellulose plastic, displayed by Parkes at Universal Exhibition, London. Austrian Army gave up guncotton as propellant, confining its use to bursting charge for loading shrapnel. Ozanam suggested use of multiple-orifice spinnerettes.

1864 Collodion emulsion photographic dry plate, Sayce and Bolton (England). Intermittent boiling and pulping in preparation of cellulose trinitrate to stabilize the product, Abel (England).

1865 Parkesine manufactured; a commercial failure.

1866 Dynamite, nitroglycerin absorbed in kieselguhr, invented by Nobel (Sweden).

1867 Dynamite improved by absorbing nitroglycerin in active base of guncotton and potassium nitrate, Abel.

1868 Celluloid, first successful pyroxylin plastic, Hyatt (U.S.). Sulphite process for wood pulp, Tilghman (U.S.).

1869 Cellulose acetate prepared by Schutzenberger and Naudin (France).

1872 Celluloid Mfg. Co. moved from Albany, N. Y., to Newark, N. J.

1875 Xylonite, nitrocellulose plastic, commercialized by Daniel Spill (England).

1878 Carbon filament for incandescent electric lamp, Swan (England). Celluloid Co. engaged its first chemist, Frank Vanderpoel.

1881 American Xylonite Co. organized, L. L. Brown, North Adams, Mass. Pasbosene, pyroxylin plastic, made by Merchants Mfg. Co., Newark, N. J. Cross and Bevan formed research partnership and cellulose studies started.

1882 Amyl acetate employed by Stevens (U.S.) as nitrocellulose solvent; start of modern lacquer industry. Weston (U.S.) made cuprammonium electric light filament.

1883 Cellonite Co., pyroxylin plastics, formed by merger of France and Kanouse interests, Newark, N. J. Nitrocellulose monofilament for electric lights, Swan (England).

1884 First commercially successful nitrocellulose textile filament, Chardonnet (France). Artificial leather, pyroxylin with castor oil plasticizer coated on fabric, Wilson and Story (England). Camphor as pyroxylin solvent-plasticizer thrown open by U. S. Court patent decision; *Spill* vs. *Celluloid Co.* Lithoid Corp., pyroxylin plastics, started at Newburyport, Mass.

1885　Swan's nitrocellulose monofilament crocheted in doilies and shown at Exhibition of Inventors, London. Cellonite became Pyralin, and Arlington Co. formed. Sulfate process for wood pulp for kraft paper, Dahl (Germany). Solubility of cellulose in zinc chloride observed, Powell (England).

1886　Pyroxylin lacquers, using amyl acetate as solvent, made by Celluloid Varnish Co., F. Crane Chemical Co., and Krystalline Co., all of Newark, N. J.

1887　Mixed butyl and propyl acetates patented as nitrocellulose solvents, W. D. Field, Crane Co. Patent for nitrocellulose photographic film applied for by Goodwin (U.S.).

1888　Lithoid Corp. reorganized as Fiberloid Corp. Cordite, double-base smokeless powder, invented, Abel and Dewar (England).

1889　Chardonnet "artificial silk" sensation of Paris Exhibition. Henry Reichenbach, employee of George Eastman, issued patent for nitrocellulose photographic film. Ballistite, explosive of mixed nitroglycerin and nitrocellulose, Nobel (Sweden).

1890　First patent for cuprammonium rayon, Despeissis (France).

1891　Chardonnet's factory produced 125 lb daily of "artificial silk." British Army adopted Cordite as propellant.

1892　Viscose rayon process invented, Cross, Bevan, and Beadle (England). Egyptian Lacquer Mfg. Co. formed by Zeller, Dolmetsch, and Borgmeyer, Elizabeth, N. J. World output of artificial filaments passed 30,000 lb a year.

1893 Viscose Spinning Syndicate Ltd. organized to develop Cross and Bevan process for textile yarns; A. D. Little (U.S.) started experimenting with viscose and acetate processes and products.

1894 Celluloid-Zapon Co. formed by merger of Celluloid Varnish, Crane, and Zapon companies (pyroxylin lacquers and artificial leather) and moved to Stamford, Conn. Bronnert (Germany) made cellulose acetate filaments experimentally.

1895 Chardonnet's factory on profitable basis, producing 3500 lb of nitrocellulose yarn daily. Little produced acetate light-bulb filaments.

1896 Strengthening of zinc-chloride cellulose filaments by stretching observed by Dreaper (England).

1897 American Pegamoid Co., pyroxylin-coated fabrics, established at Ho-Ho-Kus, N. J.

1898 Cuprammonium yarn made commercially (Germany). Rev. Hannibal Goodwin issued U. S. Pat. 610,861, application filed 1887, for photographic film.

1899 Cellulose Products Co., backed by Spruance and Saulsbury, acquired Cross and Bevan U. S. patent rights from Little, and started unsuccessfully to make viscose rayon in U.S.

1900 Spinning box for viscose yarn invented, Topham (England). Cellulose Triacetate film made by Chemical Products Co., Waltham, Mass., A. D. Little. American Pegamoid moved to Newburgh, N. Y., and name changed to Fabrikoid Co. First commercially successful machine for spinning filaments continuously, Glanzstoffabriken A. G. Cuprammonium yarn introduced into England, Bronnert.

1901 Improved process for stretch spinning cuprammonium filaments, Dr. Edmund Thiele (Germany). General Artificial Silk Co., Lansdowne, Pa., attempted unsuccessfully to commercialize viscose yarn. Tricresyl phosphate patented as pyroxylin plasticizer, Zühl and Ersemann (Germany).

1902 Cellulose acetate filament yarn experimentally spun, A. D. Little (U.S.). World annual production of rayon passed 5,000,000 lb.

1904 Acetone used successfully as solvent for cellulose acetate, George Miles. Genasco Silk Works (S. W. Pettit) formed of bankrupt Cellulose Products and General Artificial Silk companies. Courtaulds purchased Cross and Bevan viscose process rights in England. Revolving spinnerette plate to give twist to rayon fibers, Société Française de la Viscose.

1905 Perry-Austin Co., pyroxylin lacquers, organized. Fiberloid plant at Newburyport, Mass., burned and rebuilt at Springfield, Mass.

1907 Doerflinger (U.S.) incorporated cellulose acetate in lacquer formulas. Rayon waste chopped and spun into thread in Germany; forerunner of staple fibers.

1908 Continuous transparent sheeting experiments, Brandenberger (France). Eastman Kodak made cellulose triacetate photographic film with trichloroethylene solvent. Camille and Henri Dreyfus started cellulose acetate researches. First Italian rayon factory established.

1909 Courtaulds bought U.S. rights to Cross and Bevan viscose patents from Pettit and organized Viscose Company of America, Marcus Hook, Pa. Kodak adopted acetone as solvent for acetate photographic film. Mixed rayon-cotton fabrics introduced as linings (England) and neckties knitted of rayon became popular here and abroad. Value of cellulose plastics sold in U.S. passed $5,000,000.

1910 Dreyfus brothers made cellulose acetate photographic film at Basel, Switzerland. Du Pont acquired Fabrikoid, artificial leather.

1911 Viscose Co. of America (now American Viscose Co.) first year produced 362,544 lb of yarn. *The Nitrocellulose Industry,* by E. C. Worden, published.

1912 Continuous process for casting cellophane perfected, Brandenberger (France). Cellulose ethers investigated as plastics, H. Dreyfus (France), Leuchs (Germany), Lilienfeld (Austria). Lustron Co., triacetate fibers, organized by A. D. Little.

1913 Viscose sausage casings invented, Cohoe (U.S.). World production of rayon reached 25,000,-000 lb yearly.

1914 War demand for nonflammable coating for airplane wings was first big demand for cellulose acetate. Perkin Medal awarded to Chardonnet. Lustron made triacetate yarn in modest quantities. Value of U.S. cellulose plastics reached $9,000,000.

1915 Du Pont acquired Arlington Co., pyroxylin plastics. Dreyfus brothers built and operated acetate dope plant in England for British Air Force.

1916 First knitted fabric for outerwear made from rayon (Germany). Second American Viscose plant built at Roanoke, Va. Celluloid Co. enlarged plant at Newark, N. J. Industrial Fiber Co. (now Industrial Rayon) organized. Du Pont combined Fabrikoid operations with Fairfield Rubber Co. at Fairfield, Conn. Celluloid Co. planted camphor trees in Florida.

1917 Dreyfus brothers started building plant at Cumberland,

Md., to make acetate dope for U.S. Army.

1918 Visking Corp. organized to make sausage casings, Cohoe patents. World annual output of rayon, 35,000,000 lb.

1919 Celanese, first commercially successful acetate yarn, offered in England by Dreyfus. Semi-automatic compression molding presses for plastics introduced. U.S. exports of cellulosic plastics this year exceeded $8,000,000, five times prewar.

1920 Basic viscose patents in U.S. expired; du Pont and Tubize built plants at Buffalo and Hopewell, respectively; third American Viscose plant, Lewistown, Pa. Price of viscose rayon declined from $6.00 a lb in April to $2.50 in October. Stamsocott Co. formed by East St. Louis, American, and Southern Cotton Oil companies to buy surplus du Pont-operated war plant at Hopewell, Va., and purify cotton

linters. Tennessee Eastman Corp. organized by Kodak to make acetic acid and acetone, later acetate film and plastics. Celluloid and du Pont abandoned Florida camphor plantations.

1921 Southern Chemical Cotton Co., Memphis, Tenn., organized by Mercer Reynolds. Refined tricresyl phosphate marketed, Celluloid Co. World production rayon passed 60,000,000 lb.

1922 Dissolving pulp of above 90 percent alpha cellulose on market, Brown Co.

1923 First automobile, General Motors' Oakland, finished with nitrocellulose lacquer. Virginia Cellulose Co. bought Stamsocott cotton linters purification business.

1924 Rayon adopted as generic term for "artificial silk." Celanese, acetate yarn, made in America. Du Pont Cellophane Co. in operation, Brandenberger patents, Buffalo, N. Y. Acme Rayon

Co. in production at Cleveland, O.

1925 Eastman Kodak Research Laboratory, Rochester, N. Y., opened. Control of Tubize stock bought by R. L. Taylor and associates. Belamose and Skenandoa Rayon companies organized at Rocky Hill, Conn., and Utica, N. Y.

1926 Hercules Powder Co. acquired Virginia Cellulose Co. First and only cuprammonium rayon plant in U.S., American Bemberg Corp., opened at Johnson City, Tenn. Semidull viscose yarn, American Viscose Co., and first crease-resistant process patents, Tootal-Broadhurst (England). Delaware and Amoskeag rayon companies formed; Industrial Rayon Co. reorganized.

1927 Celanese acquired Celluloid. Low-viscosity nitrocellulose lacquer patents, E. M. Flaherty, assigned to du Pont. Cellulose acetate in sheets, rods, tubes, offered by Celluloid and Eastman companies.

1928 First American rayon units of Italian Chatillon at Rome, Ga., and Dutch Enka at Asheville, N. C., in production. Lumarith, acetate transparent sheeting, Celanese Co. Stretch spinning of viscose yarn, Griffin, American Viscose, and Bradshaw, du Pont. World production of rayon, 350,-000,000 lb. Moisture-proof cellophane perfected by Charch of du Pont.

1929 Chardonize, first permanent dull-luster rayon by Tubize. Acetate molding powder offered by Celluloid (Lumarith) and Tennessee Eastman (Tenite I). Du Pont bought out French minority shareholders in du Pont Cellophane Co.; new viscose plant at Old Hickory, Tenn.; began acetate production at Waynesboro, Va. Sylphrap, viscose transparent sheeting on market from Sylvania Products Co., Fredericksburg, Va. New Bradford and Woonsocket Rayon companies organized.

1930 Kodapak, acetate transparent sheeting, offered by Eastman Kodak Co. American Viscose entered acetate fiber field. Third du Pont Cellophane plant, Ampthill, near Richmond, Va. Fiberloid bought by Monsanto. Dissolving wood pulp produced on West Coast, Rayonier. Tubize and Chatillon companies merged. Kodak first produced yarn. Price of viscose rayon down to under $1.00 a lb.

1931 Acetate transparent wrapping sheeting made and marketed, Celanese. Furness continuous spinning operation for cuprammonium yarn. Flaherty low-viscosity patents licensed to lacquer makers.

1932 Hercose, cellulose acetate butyrate plastic, introduced by Hercules Powder Co. Capacity of du Pont Cellophane plant at Buffalo doubled. U.S. rayon annual output of 194,000,000 lb greatest in world.

1933 Cordura, high-tenacity rayon,

introduced by du Pont. Du Pont Cellophane plant, Richmond, doubled.

1934 Tubize, only nitrocellulose rayon plant in U.S., closed.

1935 Ethyl cellulose made in U.S., Hercules Powder Co.

1936 World output of rayon passed 1,000,000,000 lb.

1937 Hercose, AP, cellulose acetate propionate, Hercules Powder Co. Ethocel, ethyl cellulose plastic, Dow Chemical Co.

1938 Fortisan, high-tenacity yarn, introduced by Celanese. Continuous viscose spinning plant, Industrial Rayon Co., Painesville, O. Rayon price down to 50¢ a lb. Fully automatic compression molding machines for plastics perfected.

1939 Methocel, methyl cellulose, introduced, Dow. Dissolving wood pulp produced at Fernandina, Fla., first in South, Rayonier, Inc.

1940 During decade of the Great Depression, U.S. output of rayon,

nitrocellulose lacquers, and pyroxylin plastics all more than
doubled.

1941 Courtaulds stock in American Viscose Corp. sold, $40,000,-
000.

1944 Sodium carboxymethylcellu-
lose (soluble cellulose gum) introduced in U.S. commercially.

1945 Celanese Corp. opened
chemical plant at Bishop, Tex.
Beaunit Mills acquired Skenandoa Rayon Co.

1946 Sylvania Corp., cellophane
manufacturer, merged with
American Viscose Co. U.S. rayon
output passed 1,000,000,000 lb a
year.

1949 North American and American Bemberg companies acquired

by Beaunit Mills from Alien
Property Custodian.

1950 Celanese built dissolving
pulp plant in British Columbia.
Olin Industries bought Ecusta
Paper Corp. and announced entry
into transparent wrapping field
at Pisgah Forest, N. C. Hercules
Powder Co. opened plant at Kla-
math Falls, Ore., to study uses of
Western woods. Dissolving pulp
from mixed pine and hardwood
by sulfite process perfected by International Paper and West Virginia Pulp and Paper companies.
U.S. rayon production for second
year (1948) exceeded 1,000,000,-
000 lb.

1971 Celanese ended its production of cellulose acetate.

3

Baekeland's Phenolic Resin, The First Synthetic Plastics

In presenting the Sir William Perkin Medal in 1916 to Dr. Leo Hendrick Baekeland, Dr. C. F. Chandler said, "When phenol is let to react with formaldehyde under ordinary circumstances, almost anything can happen but the formation of Bakelite." Dr. Baekeland's search for the right circumstances may be likened to the search of the Curies for radium. Each envisioned the end result of the work and went about the task in a methodical and painstaking way to achieve the goal.

Many brilliant and capable researchers ahead of Baekeland had tried to solve the phenolic combination. Bayer and others had sought this answer while Hyatt was working on the Celluloid combination. Kleeburg, in the 1890s, arrived at a paste that hardened quickly and was discarded because he expected a crystal. He came close, but quit too soon. Smith, Luft, Bayer, Kleeburg, Fayolle, DeLaire, Lederer, Mannasse, Speyer, Grognot, Helm, Knoll, and Lebach all tried to combine phenol and formaldehyde ahead of Baekeland and failed to come up with a useful product.

Dr. Baekeland (fig. 3–1) at the age of thirty-five had already achieved great success as a research chemist. In seeking new worlds to conquer he dedicated himself to solving the phenolic resin enigma. A small laboratory was built in Yonkers, New York, adjacent to his home, and he devoted four years to solving this widely recognized problem. The early phases of his work led him to separate the fusible, soluble resins which he called Novolaks. He wanted a resin that would not soften again once the chemical reaction had been completed.

After several years of work, he discovered the value of hexamethylenetetramine as a catalyst and the need for pressure to stop the foaming; after endless experiments, he wound up with a clear amber phenolic resin. Here was the first truly synthetic resin and the start of the synthetic plastics. The materials before Hyatt were the natural plastics; Celluloid was a semisynthetic plastics material. Dr. Baekeland's invention opened the door to the entire field of synthetics and sparked the greater growth and expansion of plastics from the novelty into the industrial fields.

Dr. Baekeland said in his own words, "How did I happen to strike such an interesting subject as that of the synthetic resins? I can readily answer that I did not strike it haphazardly; I looked for just such an interesting subject for a number of years until I found it among the many lines of research which I undertook in my laboratory." Previously he had developed the revolutionary Velox photographic printing paper for making prints at high speed independent of weather

Figure 3–1   Dr. Leo H. Baekeland in 1916.

conditions. This invention was sold to Eastman Kodak Company for a sum that enabled Dr. Baekeland to choose the synthetic resin field as an attractive and challenging long-term research goal.

Dr. Baekeland's original aim, in his effort to find a shellac and varnish substitute, was to combine phenol and formaldehyde into a synthetic resin and make a superior varnish material that, subsequent to its application, would harden into an insoluble and infusible condition by the completion of a chemical reaction. Tremendous quantities of this resin are used today for paints and varnishes.

The following report is quoted from a talk given by Mr. R. B. Lowe in 1945:

History is a bit vague, but apparently when the early trials gave the already known unmanageable reaction, resulting in apparently worthless, blistered, insoluble, infusible lumps of resin, an assistant, Nathaniel Thurlow, became enthusiastic over the permanently fusible resinoids, while Dr. Baekeland resolved to study the heat-reactive types. It is said that he reasoned that, if he could learn to make the material do his bidding and become hard, strong, infusible, and insoluble, where and when and in what form he chose, he would produce articles of great commercial value. In the fusible types of that moment he saw, not a substitute for shellac, not a poor substitute for resin, because the fusible types had none of the toughness that character-izes shellacs. Afterward, as we know, the permanently fusible type became of equal importance through the process of hardening them with hexamethylenetetramine. Dr. Baekeland found means and methods of controlling the reactions. He could stop them at will while the resin was still in the fusible, soluble stage (which he named Stage A). He learned that, with the resin in this stage (fig. 3–2), he could dissolve it in solvents or mix it with fillers and then harden it progressively through the B stage (where the resin, although practically infusible and in-

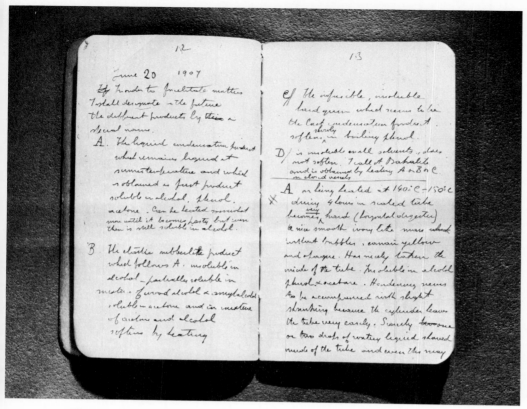

Figure 3-2   A page from the notebook of Dr. Leo Hendrick Baekeland in 1907.

soluble, was still thermosoftening) to the final insoluble, infusible C stage. The basic inventions were in formulas and techniques, which for the first time, controlled the material and resulted in usable and useful products. The classic Heat and Pressure Patent, which was and is the sine qua non of the plastics industry, exemplifies these. The so-called fifth-of-a-mole patent, relating to catalysts, is another example. On this base, application work spread into some old and many new fields. This work was covered by patents giving Dr. Baekeland a wide basic control. It is of interest to note that Dr. Baekeland's close friend and coinventor in the electrochemical field, Mr. C. P. Townsend, was also his patent attorney. Mr. Townsend was for many years associated with Carbide and was at one time head of the Union Carbide and Carbon Research Laboratories, Inc.

The early staff consisted of Dr. Baekeland, Mr. Nathaniel Thurlow, Dr. August Bothelf, and Lewis Taylor, who doubled as helper and chauffeur to the Baekeland family. Later Jim Taylor, brother of Lewis, became a

Figure 3–3   This building housed the laboratory on the Yonkers, New York, estate of Dr. Baekeland, where he developed phenolic resin.

full-time helper. With this small group and the building and equipment which you may see in figures 3–3 through 3–5, basic work, the reduction to practice and small-scale manufacture, was carried on.

The initial molded product application is credited to Richard W. Seabury* (fig. 3–6), who tells this story:

In 1906, I was running a small factory

* Richard W. Seabury envisioned an even

in Boonton, N.J., called Loando Hard Rubber Company. This plant reclaimed rubber from old bicycle tires and sold about 10 tons per month to American Hard Rubber Company, Western Electric Company, and others. The balance of the output, in the form of a rubber solution, was mixed with asbestos fibre and molded

greater future in electronics as a result of his contacts with wireless makers and in 1922 founded the Radio Frequency Laboratories and in 1928 the Aircraft Radio Corporation, from which he retired in 1951. Boonton Rubber Company was reorganized into Tech Art Plastics Company in the interim.

Figure 3–4 *Interior view of Dr. Baekeland's Yonkers, New York, laboratory.*

Figure 3–5 *Gas-heated rolls were used by Dr. Baekeland in his early compounding work.*

Figure 3–6   These pioneer molders of plastics were members of the Boonton Rubber Company baseball team. Richard W. Seabury, President of Boonton Rubber Company, was an ardent baseball fan and always a winner. (He is shown at the left end of the center row.) He was first to envision the use of Dr. Baekeland's phenolic varnish as a binder for molded products and became the first molder of Bakelite. He obtained varnish from Dr. Baekeland and made an asbestos compound which he molded for Weston Instrument Company. Dick Seabury later founded High Frequency Laboratories and died in 1970.

Figure 3–7   These bobbin ends were the first molded Bakelite parts. They were produced of a phenolic asbestos combination by Boonton Rubber Company for Weston Electrical Instrument Company.

in cast iron molds and sold to all the large electrical manufacturing companies under the trade name of Vulcanized Asbestos. About this time, the Vulcanized Rubber Company of Trenton wanted us to try to reduce the inorganic material in our Loando rubber and we consulted with a chemist, Dr. Leo H. Baekeland of Yonkers, N.Y. He was interested in a synthetic varnish. I was interested in the resin itself to take the place of rubber in molded electrical insulation. The first molded Bakelite parts [fig. 3–7], some with asbestos and some with wood

flour as the filling material, were made by me in Boonton in 1907. Dr. Baekeland brought the resin from his Yonkers laboratory, and I made the molding powders and molded the parts in the molds and hydraulic presses that I had in Boonton. A short time after this, we organized the Boonton Rubber Manufacturing Company and made a contract with Baekeland to supply the raw Bakelite resin. My first customer was the Weston Electrical Instrument Company of Newark, New Jersey. They had great need for precision bobbin ends, which could not be made satisfactorily from the rubber asbestos compounds. [A page from Richard Seabury's notebook is shown in fig. 3–8.]

In February 1909, patent coverage having established Dr. Baekeland's priority and various "quiet" trials looking promising, the doctor presented the first paper on his work, before the New York Section of the American Chemical Society, entitled "The Synthesis, Constitution and Uses of Bakelite." This announcement, coming as it did from a man who had already distinguished himself in the field of science and industry, attracted wide attention and resulted in a very real interest not only on the part of prospective users, but also prospective competitors. The doctor proposed that the material might be scientifically named oxy-benzyl-methylene-glycol anhydride. Nevertheless he registered it under the trademark as "Bakelite." Material was manufactured and sales made.

Dr. Baekeland now began to lay plans for organizing a company to exploit his inventions. Formaldehyde was a critical and important mate-

C. A. 222

Standard

82 lb   C. Resin } Grind together
 8 "   Heya   }   add
46 "   "C" Fibre
30 "   Fine Canvas
30 "   Coarse "
 4 "   Carnauba Wax
─────
200 "

Mix Well in Forge Ball Mill

erial. The sole producer in this country was Roessler & Hasslacher Chemical Company. Dr. Baekeland had a most profound respect for Mr. Jacob Hasslacher. Roessler and Hasslacher, on the other hand, were looking for outlets for formaldehyde and were keenly aware of the potential in Bakelite. They were well prepared in every way to furnish the type of support for the enterprise that Dr. Baekeland judged to be necessary. Louis M. Rossi, who had been the manager of the Roessler and Hasslacher formaldehyde plant, was taken over by Dr.

Baekeland, and he ran the first Bakelite plant in Perth Amboy, N.J. Lou Rossi was the production genius that developed the action and results. Dr. Baekeland did not like the day-to-day running of the business, and he was greatly annoyed by the small decisions that had to be made. He never participated in the operation of the business as an active manager. Lou Rossi, [fig. 3–9] was fondly called the "boss" within the Bakelite organization. Messrs. Hylton Swan, Lawrence C. Byck, Harvey D. Shannon, and Gilbert L. Peakes were early employees

Figure 3–9   Louis M. Rossi was the production genius who started the commercial production of Bakelite resins.

who contributed greatly to the growth of the thermosetting molding compounds. The Peakes-Rossi flow tester was the first standard measuring device for the thermosets.

Concurrently, Dr. Baekeland desired to establish a company to exploit his inventions in Europe. To this end he approached the Rutgerswerke Actien Gessellschaft, a large German company that dominated the acid field. A quick deal was made, and their factory located in Ecknerberi, Berlin.

In the meantime, production and sales continued in this country. I quote here verbatim, a description of the facilities and methods as recalled by Lawrence Byck.

By the middle of 1910, it was all too obvious that Bakelite had outgrown its production facilities which we dignified by the name, pilot plant. This, by the way, consisted of one small cast iron jacketed still [fig. 3–10], little "Old Faithful." Old Faithful, complete with reflux and distilling condenser and receiving tank, was set up in a corner by Dr. Baekeland's garage,

Figure 3–10   Shown here is Dr. Baekeland's first production still that was used in the developmental work which followed his invention of phenolic plastics.

adjacent to the laboratory. After one serious laboratory fire, the doctor decided that he would rather lose the garage than the laboratory. Old Faithful had an agitator which required motive power, and electrical lines had not yet reached Harmony Park. The doctor acquired a steam engine from an old White steam automobile, rumor having it that this came from one of his own early cars. This little engine, through a noisy chain drive, provided a more or less dependable source of agitation. It is still an unanswered question whether or not the early batches of "B" were due to the agitator stopping or whether the agitator stopped because the contents of the still went to "B"!

Old Faithful and its steam engine required steam at about 80-lb pressure so a small coal-fired boiler had been set up in one corner of the laboratory and the steam was piped across to the garage. This boiler required a licensed engineer to operate it so, Mr. Thurlow was induced to become the licensed stationary engineer.

A still must be charged with raw materials. This was easy. Fortunately, the major bulk raw materials in use at that time, cresol and formaldehyde, were both liquids. A hand-operated gear pump on tripod legs easily straddled a barrel of raw material standing on a platform scale.

Just a modicum of hand labor was required to transfer from barrel to the small manhole of the still. In making the first varnishes, addition of the alcohol at the crucial moment had to be made much more quickly than was possible with the little hand pump. So the alcohol was dumped onto the hot resin through the open manhole by hand from buckets. This was always an interesting, if not to say an exhilarating moment. Lewis Taylor did this, invariably with the entire staff (and frequently the Baekeland family) as audience—at a safe distance. Alcohol vapor fires were commonplace; you smothered them out by the simple expedient of slamming shut the manhole door, cutting off the oxygen supply. The fires frequently flashed up the condenser and started small fires in the second-story storage room of the garage.

These are the conditions under

Figure 3–11   General Bakelite Company started operations in Perth Amboy, New Jersey, in 1910.

which Bakelite was first made for sale in 1909, and, by mid-1910, it was demonstrated that here was a business with large potentialities. The electrical industry was expanding rapidly, the automotive industry was in its lusty but fast-growing infancy. Both urgently needed a new insulating material, better electrically, more heat stable, stronger, and more amenable to mass-production methods than anything then available. The laboratory was outgrown.

At this point a lease was taken on the building shown in figure 3–11, and General Bakelite Company started operations in 1911.

The first users of phenolic plastics were started by Dr. Baekeland who worked personally in many plants to solve the early problems. These first users, Boonton Rubber Company, Westinghouse Electric Company, Remy Electric Company, General Electric Company, Western Electric Company, Kellogg Switchboard Company, Albany Billiard Ball Company, Wagner Electric Company, etc., spawned many local molding shops. Most of the pioneer molders were initially toolmakers, turned moldmakers and then molders.

Typical examples are: Boonton Molding Company and Tech Art Plastics Company were offspring of Boonton Rubber Company. Albany Billiard Ball Company started Niagara Insulbake Specialty Company; Chicago Molded Products Company was the child of Plymouth Manufacturing Company. Bakelite initially limited its effort to a few large companies and tried to protect them from small competitors. Earl Howard tells the following story about how plastics came to the West Coast:

In 1916 I had a metal-working plant down at 9th and Los Angeles Sts., Los Angeles. I was making parts for Gilfillan Bros., who manufactured electrical and radio equipment. As the reader will remember, that was before factory-built radio sets were thought of. Each radio amateur built his own crystal set with parts made by Gilfillan and other firms in the same business.

About that time, Gilfillan went into the manufacture of replacement parts for ignition systems on all makes of cars: distributor heads, timers, etc. They wanted certain parts molded of Bakelite (nobody called it plastics then), but they knew nothing about molding, or about molding materials or equipment.

So I was commissioned to go East, visit molding firms and material manufacturers, get the necessary information, and come back and set up the Gilfillan molding plant.

Naturally, the first people I got in touch with were the Bakelite people —Dr. Leo Baekeland and Hylton Swan, his sales manager. I got absolutely no encouragement from them. Their output was small, they said, and was entirely taken up by big electrical firms like Delco. They were not anxious to spoil their relations with their big customers by encouraging a competitor, even if the competitor was way out on the Pacific Coast. So, at least at first, they refused to sell me any molding powder.

Well, at least I thought I would try to find out something about the technique of molding. So I went down to Dayton, Ohio, and to New York to the Berry Varnish Co., and finally to the Niagara Insulbake Co., at Niagara Falls.

In Dayton, and in Chicago, I had placed ads in the papers advertising for men who knew anything about molding or molding equipment to come to Los Angeles and help us set up the Gilfillan plant. I got replies from a lot of men who wanted to come to Los Angeles; but few of them knew anything about molding. They were willing, but what they really wanted was a free ride to Los Angeles. I was disappointed.

At Niagara Insulbake I got a cordial reception, for a change. When I asked their management for information on molding, they replied that the best way to learn was to go to work.

I did. I got into my overalls, and for two weeks I worked for Niagara Insulbake, molding Bakelite parts. At the end of that training period I thought I knew a little about molding, so I went back to New York for another try at the Bakelite people.

It took a lot of persuading, but perhaps Swan and Dr. Baekeland were impressed by the fact that I was still around after all that time. At any rate, they let me have 25 lbs. of Bakelite, which I packed in my suitcase and brought back to Los Angeles on the train.

From sketches I had made in the East and on the train homeward bound, we built our first compression presses. I had bought a hydraulic accumulator from the Burroughs Co. in Newark, N.J., and I had the Commercial Iron Works of Los Angeles build our first press. In a short period we had 14 presses built by Commercial. In the end, before Gilfillan was through with thermosetting molding, our plant had grown to 44 presses, 80-ton and 40-ton units. [*Pacific Plastics,* Dec. 1948].

Dr. Baekeland reported that his hardest job was teaching people to

Figure 3–12 The No. 1 Autographic Kodak Special Camera was the first to use plastics. The end panels were molded of Bakelite in 1914.

use this new product correctly, avoiding previous habits with shellac and rubber. He expected that anyone who understood molding would be able to make a success of Bakelite. He learned the hard way that it is difficult to teach an old dog new tricks. The knowledge of how to mold rubber, shellac, or Celluloid in many cases was a block to progress. These older materials were thermoplastic. Bakelite was hardened by heat in the mold and needed

high temperatures and high pressures to produce the desired properties. He found that many older molders were habitually timid; they worked with low heat and got bad results.

He had experienced the same problem previously with Velox paper; veteran photographers damned it because it would not work under their archaic methods. He reported that some of the most successful molders of Bakelite were those who

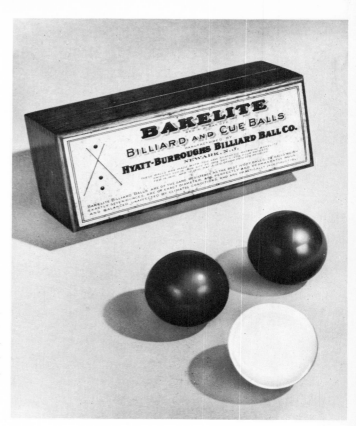

Figure 3–13   Gutta-percha had replaced ivory for billiard balls long before Hyatt's day. In 1912 the Hyatt Burroughs Billiard Ball Company changed to the use of Bakelite since it was more economical to produce and made a superior product.

had never molded the older plastics. An early product is shown in figure 3–12. It is interesting to note that the plastics industry repeated this same unfortunate stubbornness again when the thermoplastics and injection molding revolution started. The older compression molders looked with disdain on this new process also and moved in after the thermoplastic newcomers had made big progress.

Baekeland, Burroughs, and Hyatt worked together on a better billiard ball as shown in figure 3–13. In subsequent years, the Albany Billiard Ball Company that was founded by the Hyatts to make Celluloid balls, changed over to a cast phenolic resin. The inserts are cast in place and glass lamp bulbs are used for molds for the white phenolic resin casting process. A very long annealing cycle, grinding, coloring, and engraving completes the balls.

News of the phenolic resin development inspired Thomas A. Edison and his associates to initiate studies of its potential for phonograph records. His competitors were driving him away from the cylindrical record, and he wanted to offer an unbreakable flat record. Mr. J. W. Aylesworth, who was then Chief Chemist of the Edison organization, produced some fine quality resins, but they were not really suitable for the records. Messrs. Aylesworth, Dyer, and Kirk Brown, who had substantial experience in the rubber field, then formed the Condensite Company of America (fig. 3–14) to exploit Aylesworth's resin. Mr. Edison did not join this enterprise. Condensite was soon to be confronted with the Baekeland patents which would have put them out of business. Mr. Edison made a strong personal appeal to Dr. Baekeland on behalf of his associates, and an arrangement was made whereby Condensite obtained a li-

cense under the Baekeland patents and continued to operate. They pioneered the sale of compounds instead of resins and developed the two-step, hot-discharge materials which accelerated the use of phenolics tremendously.

Mr. Adolph Karpen of S. Karpen and Brothers Furniture Company of Chicago had sponsored Dr. Lawrence V. Redman to undertake research work at the University of Kansas looking toward the production of an improved furniture varnish, which had also been a major aim of Dr. Baekeland. The resins produced by Redman and his associates, Frank P. Brock and Archie J. Weith, were not satisfactory for the varnish application but were competitive with Bakelite and Condensite molding resins. Redmanol Chemical Products Company was then formed by Adolph Karpen to exploit their Redmanol resins. A patent infringement suit was brought by Baekeland against Redman and an infringement judgment obtained.

Adolph Karpen thereupon proposed to Dr. Baekeland that he purchase the Redmanol Chemical Products Company or sell General Bakelite Company to him. Dr. Baekeland refused to consider either suggestion. Adolph Karpen then quietly acquired control of Condensite Corporation of America by purchasing the Aylesworth and Dyer stocks. In this manner Redmanol Chemical Products Company also acquired a license under the Baekeland patents. This led to the formation of Bakelite Corporation in 1922 by the combination of the three resin makers. The Redmanol and Condensite names were then allowed to die out. Bakelite

Corporation became a unit of Union Carbide and Carbon Corporation in 1939.

In the Condensite operations, Kirk Brown was the leader with Aylesworth the principal technical source. Kirk Brown's three sons, Sanford, Allan, and Gordon were active contributors to its success and they later assumed leadership positions in Bakelite Corporation.

An anecdote by Saul M. Silverstein, maker of the most useful "Rogers Board," gives a personal picture of Dr. Baekeland as a negotiator:

I had been trying for more than a year to get a license to use the Redman-Cheetam patents for the beater-addition process. I was wined and dined, the Bakelite executives were too polite to say "no"; but they didn't say "yes." Finally I was lucky enough to be introduced to Dr. Leo H. Baekeland, who resolved our problem in a matter of minutes by stating in his very Flemish accent, "Vee do not vant a licensee, vee vant a partner."

Use of phenolics in the electrical industry was followed quickly by its introduction into the automotive industry. "Boss" Kettering's ignition and starting system for automobiles popularized the car, and his systems demanded molded, stable dielectric materials. These versatile synthetic resins moved quickly into all fields of industry. Dr. Baekeland recognized at an early date the potential for shellac replacement in grinding wheels and patented this procedure in 1909. Without the improved grinding wheels, the automotive industry and others could not have moved easily into mass production.

The telephone industry turned gradually from rubber and shellac to phenolic for its receivers and mouthpieces. Early wireless makers found many applications for phenolic molded and laminated plastics. The great need for thin-section, high-strength plastics spawned the laminates (chapter 9) at an early date and Bakelite-

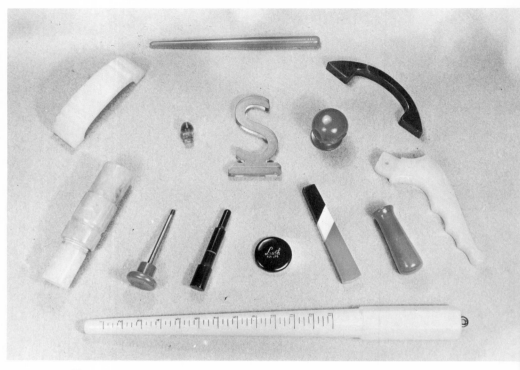

*Figure 3–15   Products fabricated of cast phenolic resin in the 1930s.
(Courtesy, Colonial Kolonite Company.)*

Micarta was developed at Westinghouse in 1911. Bakelite-Dielecto, Bakelite-Micarta, Condensite-Celeron, Formica, and GE-Textolite were pioneer entries into the field. All are produced today with phenolic and many other later resins.

Dr. Baekeland and Dr. Redman were seriously interested in a substitute for amber, and their cast phenolic resins were used by the fabricators for a variety of products, as shown in figure 3–15. A line of pipe bits and cigarette holders was sold by General Bakelite Company in the teens.

Dr. Leo Hendrick Baekeland was born in Ghent, Belgium, on November 13, 1863. He studied at the University of Ghent and was awarded a doctorate of natural science, maxima cum laude. After teaching chemistry in Bruges until

1889, he came to the United States on a traveling scholarship and decided to stay as a research chemist. Dr. Baekeland died at the age of 80 in 1944. He must be classed among the greatest inventors, and his contributions to the plastics industry are greater than all the rest.

4

Machinery for Thermosets

Charles Frederick Burroughs may well be called the pioneer plastics machinery builder and innovator of methods. Joining his father's company at the age of 18, he met John Wesley Hyatt, who had an office in the Burroughs plant. The young Burroughs worked closely with Hyatt in the development of hydraulic machinery, molds, methods, and fixtures. He designed much of the early plastics molding and fabricating equipment, with and after Hyatt. Charles Burroughs is credited with semi-automatic molding (fig. 4–1), preforming, flash molds (fig. 4–2), angle presses (fig. 4–3), tilting-head presses (fig. 4–15), and many other developments. In later years, the Burroughs Company was taken over by Watson Stillman Company. Walter Rahm, a widely known DuPont engineer, and Louis Rahm, renowned for his educational work, started their careers with the Burroughs Company.

Figure 4–1 This early semiautomatic press is typical of the presses used during World War I. (Courtesy, The Charles Burroughs Company, Newark, New Jersey.)

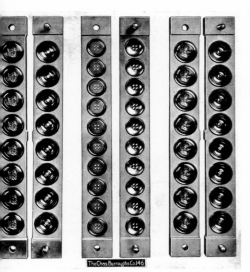

Figure 4–2 The button industry started with flash or overflow molds such as these. (Courtesy, The Charles Burroughs Company, Newark, New Jersey.)

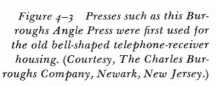

Figure 4–3 Presses such as this Burroughs Angle Press were first used for the old bell-shaped telephone-receiver housing. (Courtesy, The Charles Burroughs Company, Newark, New Jersey.)

Figure 4–4   The early hand-screw presses were replaced by hand-pumped hydraulic presses such as this early Burroughs pump. A large gain in closing speed was realized therefrom. (Courtesy, The Charles Burroughs Company, Newark, New Jersey.)

Figure 4–5   This is one of the first self-contained hydraulic presses as built for hobbing by The Charles Burroughs Company. (Courtesy, The Charles Burroughs Company, Newark, New Jersey.)

Figure 4–6 The multiple-press plants used hydraulic accumulators such as these. These accumulators are shown as they were used in 1915 at Albany Billiard Ball Company.

Other very early press builders were Charles F. Elmes Engineering Works of Chicago; Terkelsen Machine Corporation of Boston; H.P.M. of Mount Gilead, Ohio; and Commercial Iron Works of Los Angeles, California.

Hand pumps, as shown in figure 4–4, were often used for hydraulic pressure in the era between the hand-screw press (fig. 1–13) and the engine-driven hydraulic pump. An early "self-contained" unit is shown in figure 4–5. This was followed by steam or electrically driven pumps that built up pressure in an hydraulic accumulator as shown in figure 4–6. In the later years, up through World War II, the larger plants had dual accumulators as shown in figure 4–7 plus a labyrinth of pipe depicted in figure 4–8 to distribute the low and high pressure fluid. Numerous air lines further complicated the picture.

Simple quick-turn valves were

Figure 4–7   Prior to the development of self-contained presses, large pump and accumulator systems, such as this 1947 system at Shaw Insulator Company, were used for hydraulic power.

Figure 4–8   The introduction of air-operated valves permitted mechanized press operation at Shaw Insulator Company in 1946. (Courtesy, Shaw Insulator Company.)

The Chas. Burroughs Co. 112

Figure 4–9 This quick-operating valve was widely used for hydraulic pressure during the 1920s. (Courtesy, The Charles Burroughs Company, Newark, New Jersey.)

used on the early presses as shown in figure 4–9; these were followed by air-pilot operated valves in the forties to permit automatic control. Builders of injection molding machines developed self-contained pumps which were ultimately used with each individual thermosetting press in the postwar era.

Hand molds as shown in figures 4–10 and 4–11 were used for the early molding of shellac, Celluloid, and the phenolics as developed by the pioneer molders and described in chapter 1. The Burroughs hand mold press battery illustrated in figure 4–12 was extensively used by the early phenolic molders, and some are still in use today.

In 1923 Condensite Company developed compounds that permitted the molded parts to be discharged hot from the mold. Prior methods depended on compounds that had to be cooled in the mold before ejection. The advent of these hot-

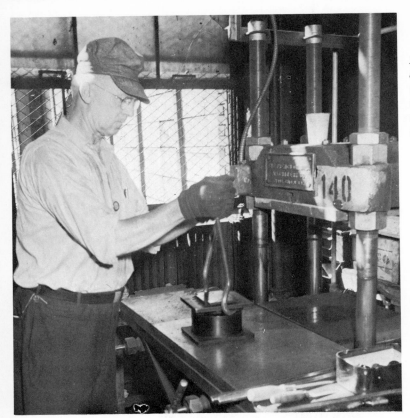

Figure 4–10   Hand molds such as this were used by the molder in the days before Hyatt, and they are still used today for some small-volume thermosetting products.

Figure 4–11   A hand mold was often a monstrous thing that required more than muscle.

Figure 4–12   One man ran this hand-mold press battery in the decade after 1910. The left front platen kept molds and materials hot during loading. The second and third presses clamped molds alternately during the long cure, and the right hand press was used for disassembly and knockout. Some of these old Burroughs press batteries are still in operation. (Courtesy, The Charles Burroughs Company, Newark, New Jersey.)

Figure 4–13   The semiautomatic mold was permanently clamped in the press and equipped with a built-in heat source and knockout system. Shown here is a typical press of the late teens. (Courtesy, Western Electric Company: "Plastic Molding as Used in the Telephone Art," Western Electric News, January 1927, page 8.)

Figure 4–14   This press battery of 1938 was designed to facilitate the operation of several presses by one operator. Single-turn valves, air knockouts, eye-level mold mounting, and fast-close rams enabled one woman operator to run up to five presses and get 200 heats per press. Her work was limited to valve opening, part pickup, blowout, remove charge from preheater and place it in mold, swing valve to close, and reload preheater. (Courtesy, General Electric Company)

*Figure 4–15 These Burroughs tilting-head presses were used in the
1920s to facilitate loading, unloading, and unscrewing parts from large
multiple-cavity molds. Telephone earcaps are being molded in this view.*

discharge compounds was another
milestone for the phenolics.

With the introduction of semi-
automatic molds (shown in fig. 4–
13) the heating system was in the
mold chaise and no longer a part of
the press. Initially one man ran one
press; economic conditions suggested dual press operation because
the long cure time permitted the
operator to change the alternate
mold. The depression years fostered
"battery" operation with one per-

son running several presses as depicted in figure 4–14.

Many special presses were developed for the compression molding
of thermosetting materials. The Burroughs Tilting-Head Press shown
in figure 4–15 was used for large
multiple-cavity molds to facilitate
removal of the part. A sliding tray
press and mold for telephone mouthpieces is shown in figure 4–16 with
an in-line clamp. The first of the
giant presses is shown in figure

Figure 4–16   This old press was designed to permit the use of sliding tray molds with a minimum of physical effort at Western Electric Company.

4–17. It was built by French Oil Machinery Company in 1934 for the General Electric Company molding of the Toledo Scale Housing. This pioneer press had a 36-in. ram, a 36-in. stroke, and clamped 1,520 tons at 3,000 psi. All six valves were machined in a solid, forged-steel block for high and low pressures. The next major change in press design followed the development of transfer molding.

Molders learned at an early date that preheated material flowed better and cured faster. Steam preheat tables (fig. 4–18) were first used to warm the powder and preforms. The operator often placed the preforms on the side of the mold to warm up for the next cycle. Rotary cans (fig. 4–19) in a heated atmosphere were used to preheat powder charges. Infrared lamps (fig. 4–20) provided better preheat for preforms; laminated preforms were widely used. A steam-jacketed churn

Figure 4–17 *This historic press was built to mold the Toledo Scale housing in 1934. It produced the first of the big pieces and pioneered many other developments. Harry E. Hire of General Electric Company and James L. Rogers of Toledo Synthetic Plastics Company (Plaskon) were responsible for this important work. (Courtesy, General Electric Company.)*

Figure 4–18 *A preheater of 1915. The first preheaters were called stock-heating plates and warmed the preforms. A piece of cloth was often used to minimize contact area heat. (Courtesy, The Charles Burroughs Company, Newark, New Jersey.)*

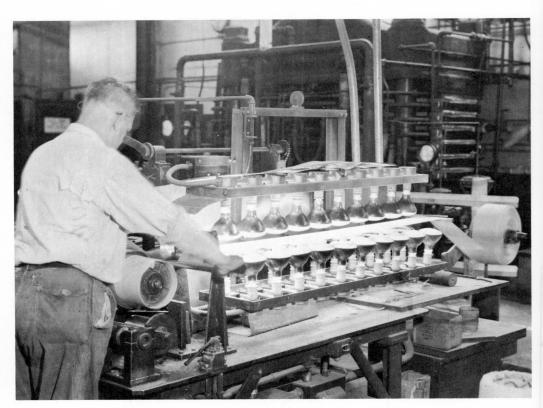

Figure 4-20 Preheating of thermosetting compounds was often done under infrared lamps. Shown here is the preheating of resin-impregnated fabric used for molded laminates.

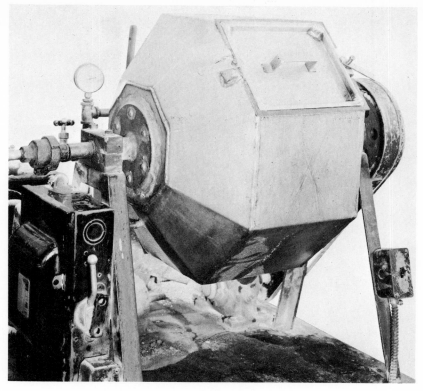

Figure 4–21   The large charge of urea powder used for molding the famous Toledo Scale housing used this steam-jacketed churn to preheat the individual load between cycles.

(fig. 4–21) was used for the large charge of urea powder required to mold the Toledo Scale housing. The advent of dielectric preheating revolutionized thermosetting molding and made obsolete most of the older preheating methods.

Dielectric or high frequency preheating (fig. 4–22) was introduced in 1945 by Bakelite's Virgil E. Meharg and an electronics consultant, Paul D. Zottu; they were given the 1946 Hyatt Award for this most important contribution to the art of molding thermosetting materials. By combining high-frequency preheat with motion study and mechanized valve control, thermosetting labor was reduced to gathering the parts, blowing, loading, and closing.

Clinton Bateholts, president of Specialty Insulation Manufacturing Company learned of Dr. Meharg's demonstration and talked about it to W. T. LaRose who was buying

Figure 4–22   This early preheater of the Radio Corporation of America is typical of many early models. (Courtesy, RCA Manufacturing Company, Incorporated, Camden, New Jersey.)

Figure 4–23   Steam preheating was found to be advantageous for some jobs. Live steam was bled into this cabinet to warm the powder by contact as it rotated in the cans. (Courtesy, Moxness Company.)

molded parts from him for General Electric Company. A formal request was made to General Electric Company via La Rose for the building of such a device. At that time, the GE Industrial Heating Division reported that such units would be too expensive to compete with steam tables, infrared, and other contemporary units. Mr. Bateholts was determined to have one regardless of its cost, and Bill LaRose offered to build it at home in his spare time. He did this, and it put him into the high-frequency preheater manufacturing business. On the first production run, this preheater reduced the cure time from 42 minutes to 6 minutes—which is significant of the great gains for plastics achieved by the use of high frequency. This sketch is typical of the foresight and will for accomplishment that built many facets of this industry.

A large demand developed immediately for the dielectric pre-heaters as a result of the tremendous cost savings that were achieved and the improved quality of the molded pieces. In combination with transfer molding, this became the most economical way to mold thermosets. Radio Corporation of America, General Electric, Airtronics, Thermall, Girdler, International Telephone and Telegraph, Westinghouse, and others all jumped into the act to supply this new market. Moxness introduced the use of steam preheat in 1948 for transfer molding (fig. 4–23).

An interesting early mold-heating method is reported for 1918 at Stokes Rubber Company: Large 2-in.-thick circular blocks of steel were heated in a coal stove adjacent to the press and then placed under the mold to transmit its heat. Steam was used as a means for heating the

molds. The presses of the Civil War era had steam-heated platens (chap. 1). Gas burners also were used for mold heating in many plants (fig. 4–24). Some of the larger plants used superheated hot water at 370° F. for mold and platen heating.

The use of preforms was slow in developing but, once started, grew very fast. Initially, molding compound was weighed by hand in small balances or in the Stokes Bakelite Measuring Machines, which discharged an exact weight into paper cups.

Tableting is reported to have started in 1922. One of the early tableting operations was recorded at Remy Electric Company for the making of coil ends. "They weighed the compound into a biscuit tin, poured it through a funnel into a preform mold, and compressed it in a hand-operated knockout press

Figure 4–25   The early Stokes Measuring Machines for dispensing weighed powder loads were followed by this automatic preforming machine that was extensively used by compression molders. With slight modification, this machine was used for automatic cold molding.

with a production rate of 3 pills per minute." A variety of preforming machines was developed by Colton, Defiance, Kux, Stokes, Standard Machinery, Logan Engineering, and others. One of the early machines is depicted in figure 4–25. A modern tableting machine (as shown in figure 4–26) will handle up to 300 preforms per minute. Noteworthy is the fact that preforms, i.e., shaped tablets designed to distribute the charge, were essential to many compression molding operations to ensure fast close and uniformity of density. High-frequency preheat and transfer molding changed all that to round or square tablets of uniform thickness.

Automatic compression molding was first demonstrated in 1934 by Bakelite Corporation with the press shown in figure 4–27. This press was designed by G. C. Gunderson

*Figure 4–26   The need for better preform-
ing machines for difficult-to-flow materials
(such as the mica-filled melamines, glass-
bonded mica, or asbestos compounds)
brought the development of this Logan Engineering Company all-hydraulic preform
machine. (Courtesy, Molecular Dielectrics,
Incorporated.)*

*Figure 4–27   This first fully automatic com-
pression molding press was designed by G. C.
Gunderson for Bakelite Corporation and
started the trend to mechanized molding.*

Figure 4–28   The Gordon Sayre fully automatic molding machine compression molded many bottle caps and other simple shapes in a fully automatic cycle that included unloading drum, preforming, preheating, molding, tumbling for flash removal, and sorting. (Courtesy, Boonton Molding Company.)

and exhibited at A Century of Progress, the 1933 World's Fair.

A completely coordinated mechanism from material in the drum to finished pieces was developed by Gordon Thayer for Boonton Molding Company. This machine, as depicted in figure 4–28, emptied the drum, made preforms, molded, ejected, and tumbled the parts. This was followed by a series of fine developments by Stokes (fig. 4–29),

Lauterbach (fig. 4–30), General Electric Company, Cropp, Watson Stillman Company, and New England Butt Company (fig. 4–31).

Transfer Molding

Looking backward, it is probable that transfer molding might have been used initially as the logical procedure for thermosetting com-

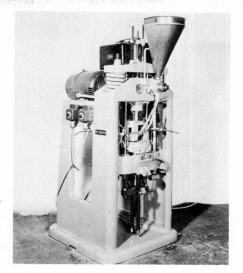

Figure 4–29   The first fully automatic molding machine that was commercially available was built by Stokes in 1936 and opened many low-cost markets to the phenolics.

Figure 4–30   The Lauterbach Press is particularly suitable for closures and other small thermosetting parts. Individual molds in a series are actuated by separate cylinders, are automatically loaded and cured during the period of rotation, and then are ejected. (Courtesy, Auburn Button Company.)

Figure 4–31   Fully automatic multiple-station compression presses followed the Lauterbach machine for high-volume simple items. (Courtesy, New England Butt Company.)

pounds had it not been for the widespread use of compression molds in the era before Hyatt and Baekeland. By the transfer molding process, all of the molder's skill was performed automatically in sequence by the mold action. Proper placement of charge, rate of close, gassing, plasticization before flow, and complete material densification are done automatically in sequence by the transfer process. Transfer molding is similar to injection molding, and the term initially applied only to thermosetting compounds. Injection molding was the term used that applied only to thermoplastics when Louis Shaw invented the closed-mold method for processing thermosets. Because the machines are similar, fully automatic transfer molding of thermosets is often called injection molding today also.

Louis E. Shaw, a moldmaker at Shaw Insulator Company, invented

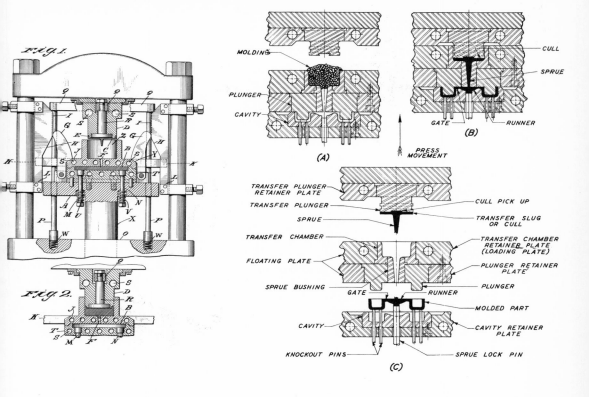

FIG. 1.

FIG. 2.

MOLDING

PLUNGER

CAVITY

(A)

CULL

SPRUE

GATE — RUNNER

(B)

PRESS MOVEMENT

TRANSFER PLUNGER RETAINER PLATE

TRANSFER PLUNGER

SPRUE

TRANSFER CHAMBER

FLOATING PLATE

SPRUE BUSHING

GATE — RUNNER

CAVITY

KNOCKOUT PINS

CULL PICK UP

TRANSFER SLUG OR CULL

TRANSFER CHAMBER RETAINER PLATE (LOADING PLATE)

PLUNGER RETAINER PLATE

PLUNGER

MOLDED PART

CAVITY RETAINER PLATE

SPRUE LOCK PIN

(C)

transfer molding in 1926. Lou Shaw's invention (shown in fig. 4–32) was concerned with difficult products, fragile inserts, intricate sections, multiple holes, and side coring of openings. The integral transfer mold served this area very well. It made possible the siphon head (fig. 4–33) that paved the way for the famous M-52 fuze shown in figure 4–34, which was produced in high volume by many molders during World War II. Shaw Insulator

Company licensed many molders to use the transfer molding process, and Frank H. Shaw, its president, was given the Hyatt Award in 1943 for his promotional work.

With the advent of dielectric preheaters, the auxiliary ram or plunger transfer mold shown in figure 4–35 became the low-cost molding procedure for most thermosetting products. This mold design was pioneered by Western Electric engineers at their Kearny Works.

Figure 4–33   The prewar soda siphon head sparked the design of the M-52 fuze in World War II. This part could not have been produced by the older compression molding. (Courtesy, Shaw Insulator Company.)

Figure 4–34   This M-52 trench mortar fuze was achieved by the transfer-molding process. Many molders produced this part in high volume, and it did much to expand the understanding and use of transfer molding for the thermosets.

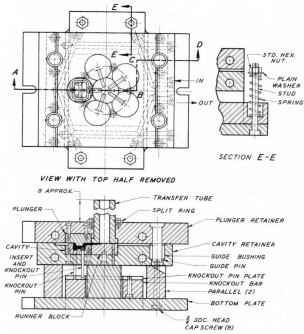

Figure 4–35   This pioneer auxiliary ram transfer mold was inaugurated by Western Electric Company. It revolutionized thermosetting molding.

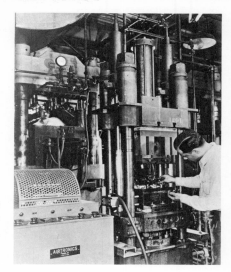

Figure 4–36   Many compression molders added a top ram to old compression presses to gain the advantages of plunger molding.

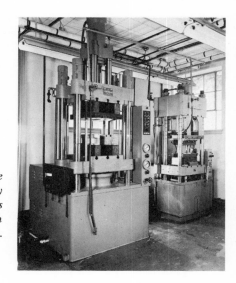

Figure 4–37   The immediate acceptance of the auxiliary-plunger transfer-molding system led many pressmakers to develop self-contained units such as these 1948 models by Watson Stillman and French Oil Mill Machinery Company.

Figure 4–38   The metal-short early years of World War II created a demand for molding presses that could not be met by conventional sources. Many molders had to build their own presses out of scrap materials. This home-made press built by V. F. Rogers produced war materials of very fine quality.

Figure 4–39   Telephone handsets were molded in this side-ram press with electronic preheat in World War II at Shaw Insulator Company. (Courtesy, Shaw Insulator Company.)

Many old compression presses were converted to the auxiliary ram type as shown in figure 4–36. The pressmakers soon offered fully self-contained high-speed presses for transfer molding with auxiliary rams and clock control as shown in figure 4–37.

Molders were very ingenious in their press conversion. The conversions took place during World War II when materials were scarce. Many presses were built on the spot with available raw materials, as illustrated by figure 4–38. The angle press shown in figure 4–39 was a homemade conversion job to produce the Navy's telephone handsets. The gunstock press (fig. 4–40) designed by Clinton Bateholts illustrates well the Yankee ingenuity that was displayed in many places where production had to be accomplished with make-do machines.

James D. McDonald, a pioneer West Coast molder, built in 1935

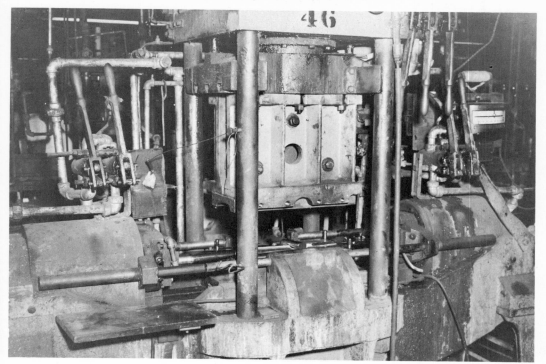

Figure 4-40 *This special press was built to mold phenolic gunstocks during World War II. It was designed by Clinton Bateholts.*

Figure 4-41 *This pioneer automatic transfer-molding machine was developed in 1935 by James D. McDonald. High-frequency preheat was added in 1943. After 35 years, it continues to run at Sierra Electric Company. It is known to have produced 3/8" square handles of phenolic on a 13-sec cycle in the early 1940s.*

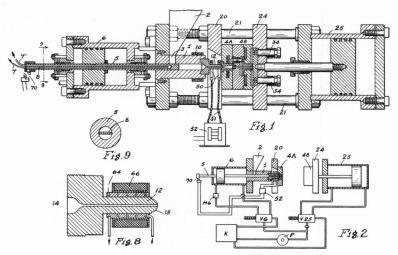

Figure 4–42 *The jet molding machine, developed by Plastics Proc-*
esses, Incorporated, in 1940, was an early injection machine for
thermosets. A conventional Reed-Prentice injection-molding machine
was equipped with a special nozzle. A high-current transformer was
used to raise the nozzle temperature almost instantly at the time of
transfer. Water cooling followed the transfer to prevent further cure
of material in the nozzle. (Courtesy, Plastics Processes, Incorporated,
Cleveland, Ohio.)

one of the first automatic transfer presses (shown in figure 4–41). Preforms were fed into the machine from the bottom of a tubular stack. High-frequency preheat was added in 1943, and the press continues to run thirty-five years later.

Jet-molding machines were developed by C. D. Shaw of Plastics Processes, Inc. in 1940. Use of this process was pioneered by Evans Winter Hebb in Detroit under the guidance of Anthony D'Agostino.

The operation is explained in figures 4–42 and 4–43. The Cousino machine followed in 1943; it was designed to handle either thermosetting or thermoplastics materials. It made use of a rotating spreader in the plasticizing chamber. Chrysler used it with limited success for rubber molding.

Fully automatic electronic preheat transfer molding was first offered by Rockford Machine Tool Company in 1946. Their machine is

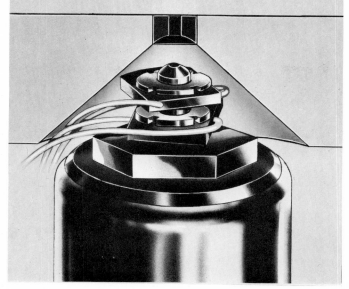

Figure 4–43 (above) Shown here is the nozzle end of the jet-molding machine. (Courtesy, Plastics Processes, Incorporated, Cleveland, Ohio.) Figure 4–44 (below) This 1947 Hyjector made by Rockford Machine Tool Company was the first fully automatic transfer-molding machine. It used electronic preheat and was 20 years ahead of its time. (Courtesy, Rockford Machine Tool Company.)

Figure 4–45   Screw plasticization with automatic transfer molding was developed by Marlin and Clyde Keaton in 1949. A nonreciprocating screw was used. The quantity of feed was regulated by a screw-revolution count control. (Courtesy, Watson Stillman Company.)

shown in figure 4–44. The Rockford Hyjector used dielectric heating and was twenty years ahead of its time. The materials of that era did not measure up to the machine's capability.

Screw plasticization with automatic transfer molding was introduced by the Keaton Brothers in 1949 and is depicted in figure 4–45. In operation, the Keaton Plungermatic Press used a nonreciprocating screw with constant flight to meter, plasticize, and push the material into the transfer tube. Quantity of feed was regulated by a screw-revolution count control that provided adequate feed for accuracy. The screw could serve as a "stuffer," moving several plasticized charges into the transfer tube in sequence before the charge was transferred into the mold by the auxiliary plunger. This is the same as loading several preforms into a transfer tube.

During this era, many companies mechanized their own transfer presses with mechanical handling of preforms or powder through a high-frequency or infrared preheat cycle and on into the transfer chamber of the mold. Delco Remy, Sylvania, General Electric, Westinghouse, Standard Oil, and many others improved their existing equipment in this manner.

At the same time Stokes, Rogers, Baker, and Hull introduced various types of automated material-handling and preheating equipment for their auxiliary plunger transfer presses to gain automatic operation. All of this effort was motivated by the need for the better properties of the thermosets at lower prices to compete with the upcoming thermoplastics.

The Rockford Hyjector and the Keaton screw transfer presses paved the way for our contemporary fully automatic thermosetting presses.

The in-line automatic plunger transfer molding process was patented by Novotny in 1944, but it was little used until George Scherry developed his automatic transfer press for Grayhill Moldtronics Incorporated.

It is noted in these historical studies that the lack of fixed opinions based on prior art often paid dividends. Dr. Baekeland reported greater progress with people who had no previous molding experience. John Slater and Spencer Palmer reported that nonmolders were more alert to the potential with thermoplastics than some of the older molders of thermosets. The development of the Grayhill automatic transfer molding machine is particularly noteworthy because of its innovation and the large gains therefrom. Contemporary molders had, in general, been satisfied with hand or semiautomatic compression molding. Grayhill could not afford what was offered, and developed its own unique answers.

The fast-changing electronics field needed low-cost precision molds and low-product cost for small-volume products. This demanded simple molds and presses, quick change, fast cycles, and low labor expense. Grayhill president, Mr. Ralph M. Hill, canvassed the industry for help and found no interest among the molders. He wisely selected in 1947 an electrical engineer, George A. Scherry, who knew little about plastics molding machines, molds, or methods. He instructed Scherry to study the basic requirements, avoid reading any books or trade papers, not to visit molding shops, and not to talk to pressmakers or to have any exposure to contemporary practice.

The initial phase of free thinking on the subject matched the basic molding sequence with the require-

Figure 4–46 This large battery of fully automatic thermosetting transfer presses produces a wide variety of products with many materials at Grayhill Moldtronics, Incorporated.

ment. The concentrated energy of George Scherry and two helpers, for three months, put together the prototype and proved its function. The finished machine has made molding history as the first fully automatic transfer molding press to be used in quantity. A significant success feature of this press is the monitoring device that locks out the press if all material is not removed with the piece and cull. Grayhill now operates 80 such machines (fig. 4–46) and has licensed Hull Corporation

(fig. 4–47) to make them for others.

Stokes introduced their "Injecto-set" machine as depicted in figure 4–48 for the fully automatic molding of thermosets in 1963, and this triggered a big resurgence in thermosets. This introduced the reciprocating screw into the thermoset field, bringing on a revolution in the molding of thermosets. This was followed by machines that would mold the thermosets by the in-line reciprocating-screw process used for

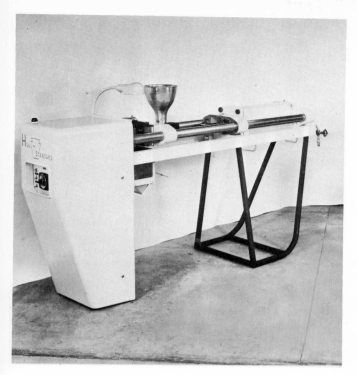

Figure 4–47   This commercial model of the George A. Scherry machine is now used by many molders.

Figure 4–48   This 50-ton fully automatic transfer molding press was introduced by the F. J. Stokes Company in 1963 and called the Injectoset. Material feeds by gravity through an opening in the barrel, which contains the reciprocating screw. Rotation advances the material along the flights of the screw as it is being plasticized by heat and mechanical shear. As the plasticized material accumulates at the end of the screw, the latter is pushed back to a predetermined stop. The transfer ram then moves to permit the screw to advance and push the plasticized charge into the transfer chamber. The plunger then transfers the material into the mold in the conventional manner. Combs are used to clear the mold and degate the parts, separating the cull and runners from molded pieces.

Figure 4–49   The cold plunger technique for automatic molding of thermosets works effectively for polyester glass and other materials that can be transfer molded at low pressure with fast cure cycles. Identical molds in series are loaded on this turntable and filled with compound at a transfer station. After a few seconds' dwell, the plunger can be withdrawn and the mold moved around the table during the cure. Ejection is accomplished at the final station, where the mold is cleaned and made ready for the next shot. This 1963 machine was developed by American Cyanamid Company.

the thermoplastics. Lewis Welding and Machine Corporation offered a fully automatic transfer press of the ram hopper-loaded type with a floating nozzle which retracted from the heated mold after the shot to minimize further cure of the incoming material. Other pressmakers then offered automatic machines for thermosets following the original Hyjector system with electronic preheating or with the reciprocating screw. American Cyanamid introduced its cold-plunger molding system in 1963 for alkyd specialty work as shown in figure 4–49. In the following years the reciprocating screw machine became the most popular method for the automatic molding of thermosets.

Molds

In chapter 1 the early work in Leom-

Figure 4–50   This unique mold was developed during the early 1920s by Western Electric Company and was the first to produce a textured surface on a vertical wall. (Courtesy, Western Electric News, January 1927, page 8.)

inster, Massachusetts, was considered. There the making of molds and fixtures was an essential part of the local industry. Standard Tool Company of Leominster pioneered many innovations in moldmaking. Other Leominister pioneer moldmakers were Guy P. Harvey and Modern Tool and Die Company. Charles F. Burroughs Company and Newark Die Company contributed much to the moldmaking practice in the East. Plymouth Manufactur-

ing Company, Western Electric Company, and Kellogg Switchboard Company were important contributors to moldmaking in the Midwest. An early complex mold developed by Western Electric Company is shown in figure 4–50. General Electric Company did much to automate compression and mold operation. Earl Howard, who started plastics molding on the West Coast for Gilfillan Brothers in 1916 and worked later with Harry

Figure 4–51   The initial beryllium copper molds made at the Gorham Company were cast from fine works of art and opened up a large market in religious artware. (Courtesy, The Gorham Company.)

H. Hahn, did much pioneer mold building and molding on the West Coast. Many molders started as mold builders.

Early molds were made of iron and steel. Louis L. Stott, who later formed the Polymer Corporation, pioneered at the Gorham Company the making of cast beryllium copper molds, as illustrated in figure 4–51. Gorham's foundry and sculptors were masters in the casting of fine bronze statuary, and they produced some highly decorative molds at reasonable cost. Beryllium copper hobs were also used for hobbing steel cavities. The Shaw Process, originated in England by Clifford and Noel Shaw for lost-wax pattern-making, established the use of ethyl silicate as a binder for refractory molds. The Avnet-Shaw Process introduced in the United States by Lionel Kavanagh is now widely used for the making of beryllium copper molds for all types of work.

Beryllium copper's better heat conductivity made it a very desirable blow-mold material.

The quick-change or unit die sets that are extensively used today were pioneered by Sven Moxness at Minneapolis Honeywell early in World War II. An example is shown in figure 4–52. During that period he introduced also laminated mold construction (fig. 4–53), stack molds, and multiple-pin unscrewing in transfer molds. Much know-how for plastics molding came from Moxness and other diecasting-trained engineers. Figure 4–54 is a mold design that greatly simplified the automatic molding of parts that required side cores. Prior to this development, such parts were produced in side ram presses or in molds that had to be removed from the press for ejection of the part. The demands of World War II expedited the development of other processing aids such as the Wheela-

Figure 4–52 (above)   Mold sets used in the first quick-change die
set for transfer molding were developed by Moxness in 1943.
Molds could be changed in 15 min in the Honeywell frame. (Courtesy. Honeywell Company.) Figure 4–53 (below)   Laminated mold
construction was used in 1944 at Honeywell for the wartime peri-
scope head molded of phenolic with very precise tolerances. (Courtesy, Honeywell Company.)

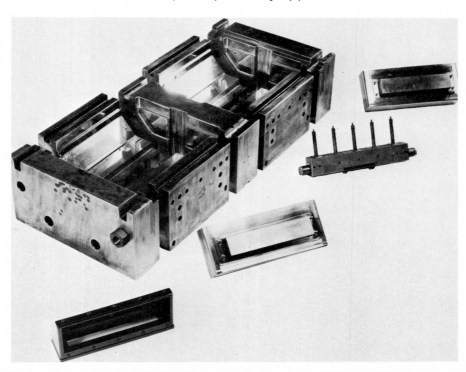

Figure 4–54   This fully automatic transfer mold greatly simplified the molding of complex parts such as iron handles. It was designed by Wayne F. Robb at Shaw Insulator Company in 1946.

brator, the Nash Rotary Lathe, standard inserts, and mold components.

Stock bases for molds were introduced by D-M-E in 1943. This cut the cost of molds considerably and greatly simplified mold design procedures. In the following years this line has been expanded to include a complete line of moldmaking components. The standard quick-change mold components for transfer molds, introduced in the early 1960s by

Master Unit Die Products Incorporated, greatly reduced mold cost and mold change time.

Lloyd Baldwin is credited with being the first to solve the mold-welding problem. Because Baldwin had welded successfully a Colton preform press, Chick Norris took a broken tool-steel Boonton mold to him for a trial at welding. Baldwin succeeded and as a result built up a very large mold-welding business. At that time it was considered

Figure 4–55   Three molds are shown here on the work bar in 1941 ready for chrome plating by the process developed by Arthur W. Logozzo. (Courtesy, General Electric Company.)

to be impossible to weld tool steel. The later development of the General Electric atomic hydrogen welders provided another major cure for the broken mold correction problem.

Chrome plating of molds (fig. 4–55) was initiated by Arthur W. Logozzo at General Electric Company in 1938. He was given the coveted Charles A. Coffin award for this development. This work was a very great boon to the molders, and today chrome plating is a standard re-

quirement for most molds. Mold wear is eliminated, better finish retained, and old molds are easily refurbished. Dimensions can be corrected in some cases and numerous other important gains resulted from Logozzo's pioneering work. He later established the Nutmeg Chrome Corporation, serving the entire country with this important specialty.

Prior to World War II, a limited amount of encapsulation had been done by casting and by transfer

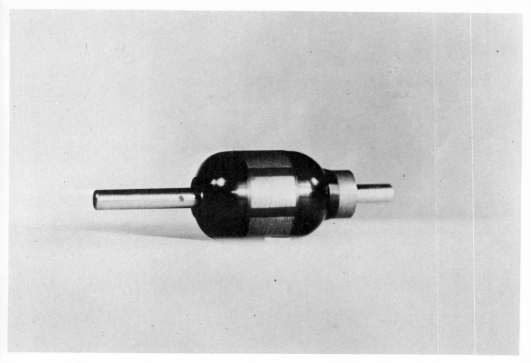

Figure 4–56 This motor armature was encapsulated with phenolic compound in the late 1930s.

molding. Motor and generator windings subject to bad environmental conditions had been encapsulated with phenolic as shown in figure 4–56. Silicone-insulated wire gave a better answer to that problem in later years.

The encapsulation of electronic components was expanded considerably in World War II for the proximity fuze. In that work, often called potting, a chassis of molded plastics was designed to hold and maintain the relationship between individual components. These parts were assembled in place, wired, and then encapsulated with a higher melt material, which did the encapsulation. Thus the spacing was maintained and uniform encapsulation achieved.

Injection and transfer presses with mold frames designed to hold the components (fig. 4–57) were developed by Hull Corp. for the electronics industry and thus provided a real practical approach to mass production encapsulation in

Figure 4–57 Hull Corporation pioneered a line of presses and mold frames with removable insert-loading frames for the encapsulation of electronic components. Shown here is the finished shot with parts runner and cull being removed from the press in the insert-holding frame. (Courtesy, Hull Corporation.)

Figure 4–58 This Capsonic fluidic control utilizes the outgoing gas to control the incoming material to avoid insert distortion. (Courtesy, Capsonic Group, Incorporated, Elgin, Illinois.)

Figure 4–59 Shown here is a polysulfide meter developed by Pyles Industries for metering, mixing, and dispensing the resin components in the proper proportion at the time that they are to be applied. (Courtesy, Pyles Industries.)

Figure 4–60 These resin tanks were used in the 1930s by General Electric Company to produce phenolic resin. (Courtesy, General Electric Company.)

the early 1960s. Capsonic Group Incorporated added another new tool (fig. 4–58) in 1969 by their development of a fluidic control that utilizes the outgoing flow of exhaust gas to control the incoming streams of plastics so that free-standing inserts are uniformly encapsulated.

Many of the more sophisticated plastics such as the epoxy and polysulphide materials must be mixed at the time of application. A variety of machines have been developed to meter, mix, and dispense such resins at the instant they are used (fig. 4–59). It is possible that instant-cure thermosets will be developed that will be mixed at the instant that they are being transfered into the mold and that will harden immediately as the components react.

A 1930 version of Dr. Baekeland's Old Faithful is shown in figure 4–60.

5

The Great Thermosetting Era

The years from 1910 to 1950 can be called the Great Thermosetting Era: moving from hand-screw presses and hot slugs of steel for mold heating to giant plants, huge presses, and electronic heating, and then along to fully automatic molding machines in 40 years. These were fascinating years of growth and technical progress.

The boundless services performed by the phenolic plastics were basic to the industrial explosion of the twentieth century. Electricity's movement into home and factory, advancing the automobile from luxury to necessity, the airplane, wireless, and radio leaned heavily on phenolic plastics. All of these things started their period of rapid growth concurrently with the development of phenolic plastics. The symbol of infinity ∞ was adopted by Dr. Baekeland, and it represented this material's infinite variety of services and applications.

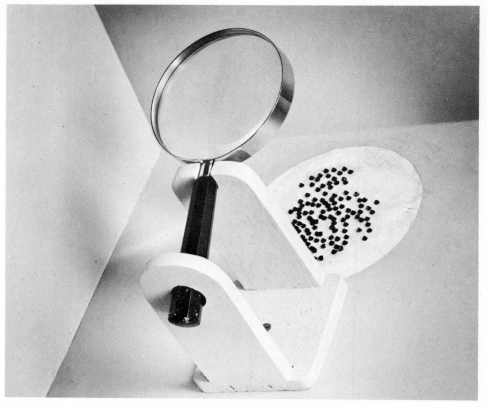

Figure 5–1   In 1910 these tiny bushings were molded of Bakelite compound by Boonton Rubber Manufacturing Company for Weston Electrical Instrument Company.

Early markets for phenolics included wiring-device insulation, instruments (fig. 5–1), commutators, components of electrical generation, distribution, and use products. Automobile distributors, magnetos, starters, and terminals used the phenolics for functional and insulating parts. Western Electric Company was soon to switch from shellac and rubber to the phenolics for telephone components. Westinghouse, working with Dr. Baekeland, developed laminated sheets (chap. 9) that became immediately basic construction materials for the electrical industry and subsequently for gears, cams, bearings, and circuitry. The decorative laminates extended into the furniture market these products that were wear and abuse resistant. Printing plates, grinding wheels, precision metal castings, automobile timing gears, radio cabinets, appliance components all moved steadily but surely onto the plastics bandwagon. The success of Dr. Baekeland encouraged others to enter this field and initiate research on plastics and resins. Condensite and Redmanol made big progress as described in chapter 3. Makalot Corporation was formed in 1924 by M. M. McKeever to produce phenolic compound. It was later sold to Interlake Corporation. Durez (Hooker Chemical Company) entered the market in 1924 after 5 years of study by Harry M. Dent, who became interested during his research on typeface materials. Durez purchased a phenol-making process in 1938 which was a great contribution to the war needs. A formaldehyde plant was purchased in 1951 by Durez. Herbert Spencer did a tremendous promotional job for Durez. Other resin makers followed after the Bakelite patents expired in 1927; many large users started to make their own phenolic resins; many already had compounding

Figure 5–2   Cast resin novelties and items for smokers were early phenolic products of special interest to Baekeland and Redman.

equipment. Resinox Corporation was formed in 1930 by Commercial Solvents Corporation and Corn Products Refining Company. George Lewis, formerly of General Plastics Company (Durez), was manager of manufacturing, and Charles Lichtenberg was sales manager. They initially produced phenolic resin for sale to those who did their own compounding. Compounds were offered in 1934, and their improved strength material

for bottle caps built sales volume that necessitated a new plant by 1938. At this point, Monsanto, who had already purchased Fiberloid Corporation, a thermoplastics maker, bought Resinox, which was combined with Fiberloid to form the Plastics Division of Monsanto Chemical Company.

An early Bakelite development and special hobby of Dr. L. V. Redman was a transparent or colored cast resin (fig. 5–2) that was used for buttons, pipe stems, billiard balls, jewelry, and, at a much later date, for radio cabinets. Liquid phenolic resin was poured in cast lead molds and cured by heat and hydrostatic pressure.

Urea Resins

The early phenolics were limited to dark colors that were not light fast, and the growing demand for color introduced the amino plastics. Frits Pollak and Kurt Ripper investigated these materials and obtained patents in 1923; Carleton Ellis did additional work resulting in urea formaledhyde patents in 1933. American Cyanamid Company introduced its "Beetle" thiourea compounds in 1928 resulting in brightly colored tumblers, buttons, and some small appliance parts.

Toledo Scales's need for a light-weight scale housing sponsored the

Figure 5–3   This historic press was built by French Oil Machinery Company and General Electric Company for the Toledo Scale housing. It was the first of the big plastics presses, having a 36-in. ram and 36-in. stroke with a molding pressure of 1,500 tons. Mold temperature was controlled by a steam compressor. This 1935 molding program started the use of plastics for business machines.

development of white molding compound by Dr. A. M. Howald at Mellon Institute in 1928, producing Plaskon urea compounds as sold by Toledo Synthetic Products Company (now Allied Chemical Company).

Molding of this largest of the plastics products was undertaken by Plaskon and General Electric Company in 1935; its success, as shown in figure 5–3, pushed all the plastics toward the fields of appliances and business machines. Need for a sturdy insulation strip between the inner and outer doors of the household refrigerator had developed the phenolic breaker strip. In time, this became the entire inner door with a white urea surface sheet, ultimately to be taken over by extruded thermoplastics after the war. This urea surface sheet answered the great need for white and pastel colors by

Figure 5–4   This 1936 urea lampshade was styled by Bill Petzold and is typical of numerous lighting applications for the plastics.

the laminates. The ureas then moved plastics into the lighting fixture fields (fig. 5–4), where a self-extinguishing translucent material was needed. Premiums, packages, closures, and buttons were large urea markets.

Melamine Resins

Liebig isolated melamine in 1834 but little was done about it for 100 years. In 1933 Palmer Griffith, working for American Cyanamid Company, was seeking derivatives of cyanamide. He produced dicyanamide and found that it contained melamine. He then added formaldehyde to a small quantity of melamine and produced a resin. Compounded with paper pulp, this resin produced a desirable molding compound.

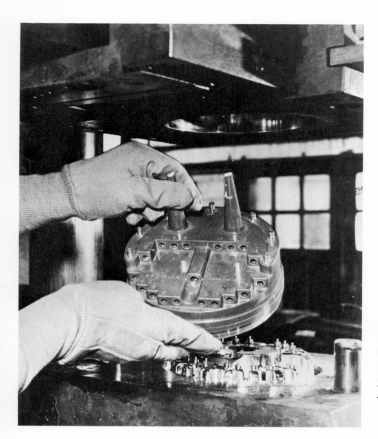

Figure 5-5 *This World War II distributor housing for aircraft was highl arc resistant and enabled our aircraft to perform at high altitudes. Melamin formaldehyde resins with mineral fille solved this problem early in the war.*

Melamine's heat resistance was superior to that of phenolics and much better than that of the ureas. Also melamine could be produced in light colors as a varnish or as a molding material or as a translucent overlay for decorative laminates. It had amazing abrasion resistance and could be used easily in place of thiourea. Its moisture resistance was superior to that of thiourea. These were badly needed properties, not previously available.

American Cyanamid initiated full scale production of melamine in 1937 and introduced its first alpha-cellulose-filled molding compound. In 1952 Griffith received the John Wesley Hyatt Award in recognition of this contribution to the plastics industry. Melamine resins brought on many gains to switchgear, aircraft distributors (fig. 5-5), and the highly successful

Figure 5–6   Cellulose-filled melamine resin created a large new market for dinnerware molded of plastics.

dinnerware programs of Boonton Molding Co. (fig. 5–6), the Navy, Prolon, and others.

The Growth Years

Dr. Baekeland's missionary and sales work, his many lectures, and the promotional work of the pioneer molders (see chapter 3) found a ready audience. The first compounds needed a half-hour cure for some small sections. The development of fast-cure compounds in 1923 by Condensite helped to accelerate the market for plastics molded products. Early applications included commutators, third-rail insulation, coil bobbins, etc. Sangamo Electric Company offered the first meter cover in 1912. An artificial hand was produced in 1914 in a mold with fusible cores.

Figure 5–7 These early Bakelite parts were made between 1909 and 1921. The distributor (upper right) was used in the World War I Liberty Motors; the Edison record was molded in 1912.

In his search for a better disc record material, Thomas A. Edison selected phenolic plastics as early as 1910 (chapter 3). At that time, it was claimed that these records (fig. 5–7) were

hard, practically unbreakable, can be played from 10 to 12 times as often as any other form of record without deterioration in the quality of their sound, and none of the quality of the original or master record is lost in the production of duplicates because of the extraordinary fidelity with which [this material] reproduces surfaces on which it is molded.

Previous records were molded of shellac compounds or wax. The vinyl compounds entered this market in 1933.

The laminated plastics (chapter 9) were developed in 1910 by Dr. Baekeland and Westinghouse. Dr. Redman and Dan O'Connor developed the early Formica materials.

Phenolic products were first introduced to the furniture field in the form of baking lacquers for brass beds. The use of these lacquers flourished, and their acceptance can best be seen from the following lines of a letter written in 1913 concerning one of these beds submerged in a water-filled warehouse during the Dayton, Ohio, flood of that year.

We . . . were surprised to find the lacquer uninjured. The finish was apparently in first class condition, all other furniture in the warehouse, outside of the bed, being completely ruined. You can imagine the disastrous effect of the muddy water. The highly polished furniture looked as though a varnish remover had been used.

The introduction in 1912 of phenolic resin coating marked a revolution in the paint and varnish industry. Previously, paint and varnish

Figure 5–8   The earpiece on this receiver was an early conversion from rubber to Bakelite.

formulas were individual secrets, painstakingly created and suitable only under certain conditions. The development of synthetic resin coatings also was in time for the many new machines and the development of new materials, which created new surfaces to be protected with paint and varnish against destructive agencies and conditions. These demands required coatings that were faster drying for mass-production line finishing and more protective for longer life.

The first phenolic package of record was a round cylindrical box with raised letters, and it was given wide publicity in 1913 as the future in packaging. Western Electric Company molded its first phone part in 1914. This was a phenolic earpiece for the old bell-shaped hard rubber receiver as shown in figure 5–8. Garson Meyer molded

Figure 5–9   Automobile ignition systems went to Bakelite insulation at a very early date.

side panels for an Eastman Kodak in 1915 of phenolic compound (fig. 3–12) on a 15-minute cycle in Eastman's pioneer evaluation of molded plastics. Eastman Kodak Company worked closely with Dr. Baekeland because of his early work for them on Velox paper.

The development of the ignition and starting system by Charles Franklin Kettering marked the entry of phenolic materials into the automotive field for, in these new plastics (fig. 5–9), automotive engineers found the dielectric strength, the immunity to adverse affects from temperatures, acids, oils, and moisture that were needed to make the automobile a reliable means of transportation. By 1916 there were more than 300,000 Delco molded-insulation sets in automotive electrical systems.

About 1916 laminated plastics

*Figure 5–10   The wireless amateurs of the late teens
did much to expand the use of phenolic laminates
and molded plastics.*

produced from phenolic resins began assuming an important place in the timing mechanism of the automobile as a timing gear (chap. 9) to start the move toward quiet operation. The year 1917 marked the development of molded phenolic steering wheels. By 1918 phenolic plastics materials had branched out to other uses such as radiator caps, gear-shift knobs, battery terminals, door-latch handles, sliding circuit connectors, commutators, and gaso-line tank covers, to be followed in rapid succession by other equally vital applications in heaters, gages, and lamp holders.

During these years the wireless amateurs (fig. 5–10) were using the phenolic laminates for tuners, variometers, tank-coil insulation, detector bases, capacitor insulation, and switch bases. They often led the commercial practice for ship-and-shore wireless transmission with their design innovation. As the con-

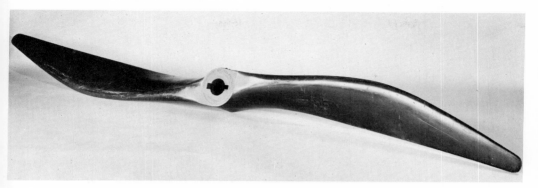

*Figure 5–11   This 10-ft propeller of World War I was molded of fabric-
filled laminated stock by Westinghouse Electric Company.*

struction of aircraft developed from the experimental stage into that of planned engineering, the future of aircraft switched from the hands of the men who flew by the seat of their pants to those who test flew aircraft on the ground. These engineers began finding durable materials that would safeguard the operation of the aircraft, which was no longer a fragile thing but a destructive as well as a constructive force. As early as 1918 phenolic plastics appeared in the ignition system (see fig. 5–7) of the famous Liberty engine as the distributor head, ignition-switch cover, and for other molded insulation parts. By 1919 distributor heads, magnetos, conductor terminals, switches, motors and generators, commutators, instrument cases, and terminal sleeves were listed among aviation parts being made of phenolic materials. In 1920 a 10-foot propeller (fig. 5–11) was molded by Westinghouse

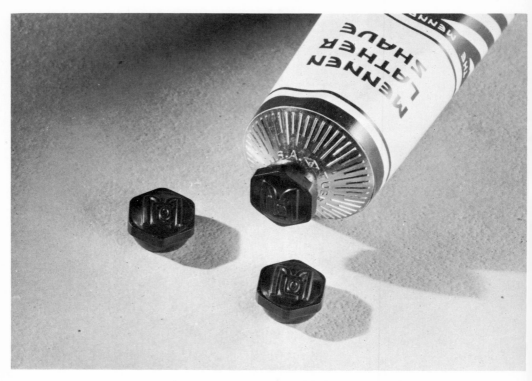

*Figure 5–12   The tin shortage of World War I converted the Mennen
cap to Bakelite in 1920, molded by Shaw Insulator Company.*

from laminates produced with phenolic resins.

World War I expanded the capacity and know-how of thermosetting molders. The Signal Corps wireless and phone sets for field use introduced demands for quality and intricacy not previously experienced. Aircraft distributors required better arc resistance and stable dielectrics. The laboratory of the battlefield pointed up the needs for better plastics and increased the understanding of environmental problems in plastics technology.

Recto Molded Products founded in 1919 by Nick Backscheider produced timers for the model-T Ford with Redmanol phenolic body and Formica T-rack. Other molded products included crystal detector bases, printing plates, and dentures with molded-in teeth.

The first caps for collapsible tubes were introduced in 1920 by the Mennen Company as the result

Figure 5–13   This early carburetor was molded of reinforced phenolic in 1923 by Shaw Insulator Company.

of the high cost of tin in the era after World War I. Sandy Brown of Bakelite Corporation developed the idea during a lunch with Bill Mennen, and these first caps (fig. 5–12) were molded by Shaw Insulator Company. For this pioneering application, the preforms were made from flat sheet. The carburetor shown in figure 5–13 was molded in 1923 complete with inserts.

Isador Leviton, founder and president of Leviton Manufacturing Company, contributed greatly to the use of phenolics for such applications as electrical switches. Starting in the button industry, he moved from the gas mantle industry to the manufacture of household electrical switches when he envisioned the total change from gas to electric lights.

In 1921, abrasives bonded with phenolic resins became a commercial reality, although they had been conceived in the early 1900s by Dr.

Leo H. Baekeland to replace shellac. Manufacturers of abrasives began using the better phenolic resin bonds, and these wheels were being applied to many different operations throughout industry. The reason for this growth was speed— speed with safety in the grinding operation. In cutoff work and rough grinding, the performance of these wheels was outstanding. They could be operated safely at 50% higher speed than shellac and vitri-fied bonded wheels; this led to the designing of new high-speed machines. Their capacity to "cut cool" with a high rate of material removal —due to the high porosity of the wheels and the strength of the resin bond—resulted not only in a saving of time and labor with a resulting decrease in grinding costs but also in an improvement of the quality of the ground product. These developments resulted in the use of resin-bonded wheels for many different types of grinding and have led to newer types of products such as abrasive belts and disks, clutch facings, and other bonded products.

Among other new materials supplied to industry were phenolic sealing solutions for porous castings. Automotive engineers at this time were finding that the porosity of an untreated cast aluminum manifold affected uniformity of the mixture and prevented the proper operation of the motor. To overcome this porosity, the interiors of manifolds and crankcases were coated during 1922 with phenolic resin enamel. This process proved to be extremely economical. Porous or leaking metal and alloy pressure castings were reclaimed by impregnating them with phenolic sealing solutions. Thus it was possible to salvage many intricate and essential castings, which might otherwise have to be rejected. These sealing solutions have been employed for

*Figure 5–14   This variocoupler, molded in brown Bakelite with lac-
quered brass fittings, upgraded the styling of radio components and
was a most popular radio receiver part in its day. It was styled by John
M. Kiefer for Adams Morgan Company, Montclair, New Jersey.*

many years and have proved to be
highly satisfactory for many types of
aluminum, nickel, bronze, and
brass castings. They soon were being employed successfully by aircraft, automotive, and electrical
equipment manufacturers, and today they are a basic tool of industry.

During 1923, phenolic resins
placed before the paint industry a
newer, more serviceable paint-
brush. This new and better bonded
brush assured painters of a brush in
which the bristles would not come
out after a short period of service
since these bristles were set in a
phenolic resin. In one dramatic test,
showing the strength of the bond, a
225-pound heavyweight was suspended in midair from the bristles.
The bond is unaffected by frequent
cleaning in water or solvents. The
introduction of these bonding resins
suggested a whole new host of applications. For example, woven and
molded brake linings were made

stronger and more resistant to wear, heat, and oil when processed with phenolic resins. Glass and mineral wool fibers later were readily formed into durable, easy-to-handle insulation bats when bonded with phenolic resins.

In 1923, the radio industry was a very large user of plastics (fig. 5–14), and Bakelite executives feared that the listening public might lose interest in radio unless the programs were upgraded. Bakelite then sponsored the first broadcast of a light opera on WEAF, which did much to attract interest to the plastics and upgrade the quality of the early programs.

The scope of contemporary synthetic plastics was widened greatly during 1924 with the development of the Weston voltmeter case. This case, molded in phenolic plastics, indicated the potential of molding plastics for housings. New products such as this opened broad new mar-

*Figure 5–15   Louis E. Shaw invented transfer molding
to produce this firing pin. (Courtesy, Shaw Insulator
Company.)*

kets. From this point on, manufacturers and product designers, who were concerned with the proper design of housings in which to encase their products, found a new answer to this problem through the use of molded plastics, not only because of the many merchandising benefits derived from the use of these materials but also because of the resulting production economies. Engineers began to realize the sales appeal and tactile properties of the

plastics, and another trend was started. It is difficult now to realize how hard it was to sell some of these obvious "firsts."

The invention of transfer molding in 1926 opened many highly technical markets (fig. 5–15) to plastics molders; this most important development is described in greater detail in chapter 4.

Multicolor molding was introduced again in 1926 when two-color contact buttons, red and green, were molded in one piece for electrical controls. Expiration of Bakelite patents in 1927 expanded the market considerably as competition entered the market. General Electric introduced its phenolic laminated breaker strips (chap. 9) for refrigeratiors in 1927 on the famous Monitor Top refrigerator.

The impact of plastics was recognized broadly by top management at this time, and competitive price reductions sparked many development with long-range implications. Gerard Swope, President of General Electric Company, foresaw that the internal plastics molding and laminating groups could not remain competitive in a captive market; these operations must be competitive when they have to sell to the open market. Harry D. Randall, who contributed much to the early work of NEMA, was given the responsibility at General Electric to develop the commercial markets for plastics

Figure 5–16   This group of men was brought together in 1931 by Harry D. Randall to form the Engineering and Marketing Department for plastics at General Electric Company. Included in this pioneer group are H. D. Randall, Roy E. Coleman, Henry M. Richardson, Frank Groten, Tom Giblin, Edw. G. Gray, Arthur Briggs, E. W. Falk, and J. H. DuBois.

in 1925, and plastics specialists were added to the Industrial Sales Department. The Plastics Department of General Electric (fig. 5–16), formed in 1931, developed training programs for engineers, and alumni of these programs are found throughout the plastics industry. Glyptal, the first of the alkyd plastics, was introduced in 1926 by General Electric Company, and it took over a big part of the coating market. The development work was done by A. McKay Gifford, Michael Callahan, and Roy H. Kienle. The Glyptals were used briefly for highly colored molded products such as the Autopoint pencil.

Studies disclosed that aluminum agitators for washing machines roughened and were not completely satisfactory. Bakelite built an experimental mold (fig. 5–17) and delivered pieces to 53 washing machine makers for test. Its 1927 introduction by Meadows Company created another very large market. Meadows patented their "use" based on less lint and less clothes destruction. By 1939 all washing machines had phenolic agitators and 120 molds were in operation on that project. Red Brannon of Bakelite sparked this development; other material makers contributed to this market expansion.

The Simmons Company of Racine, Wisconsin, in 1927 decided to explore the potential of plastics for furniture and thus became the first in that field. Mr. E. O. Wokeck, who had started in plastics at the Belden Manufacturing Company in 1914, was employed to develop the occasional chair shown in figure 5–18. It required eight separate molds at a cost of $50,000. They used a green cloth-filled phenolic compound for high strength. Wood was used for the seat to facilitate upholstery attachment. Five of these chairs were market tested, and their

Figure 5-17 This Bakelite impeller for the Meadows washing machine replaced aluminum in 1927 and made a big step in expanding the markets for plastics.

Figure 5-18 First plastics chair molded for the Simmons Company (1927-1928) of cloth-filled phenolic compound. (Courtesy, E. O. Wokeck.)

Figure 5–19 Radio tube bases were molded of phenolic plastics in completely automated plants in the late 1920s. (Courtesy, Markem Machine Company, Keene, New Hampshire.)

weight was the principle negative factor. A molded bed end was also sampled. Ladies' heels of molded phenolic were produced in 1927, but the impossible nailing problems stopped these studies until 1939, when the nailing problem was solved by the use of cellulose acetate. Many of the early failures reappeared as suitable materials were evolved. The 1929 Poole Clock changed clock cases from metal and wood to plastics almost overnight. As the 1920s were ending, the decorative applications of laminated plastics were beginning. By that year, the use of these plastics had extended into home design. Advertisements of 1929 showed 15 applications in the home for phenolic laminates, including drainboards on sinks, baseboards, fireplace facings, mantel trim, wainscoating, ceiling paneling, and kickplates for

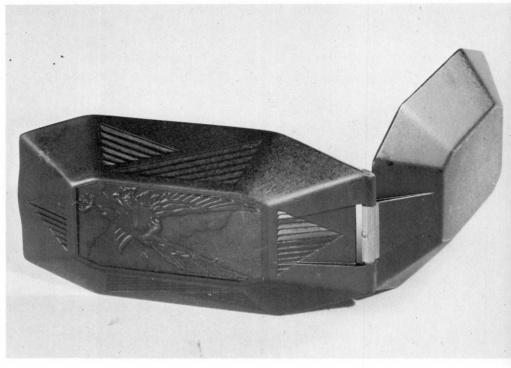

*Figure 5–20   This 1931 Elgin watch display case was molded of phe-
nolic, gold painted, and had a reversible hinged cover to facilitate
display in horizontal or vertical positions. It was produced by Chicago
Molded Products Company.*

steps. Elsewhere during 1929, this material was being used for counter plates in banks, fountain tops, and table tops in restaurants and cafeterias. Westinghouse, Formica, and General Electric Company started promotional work to increase the use of plastics in the home (see chap. 9). The great days of radio were in full swing, and beautiful as well as functional molded products caught the public fancy.

The second decade of thermoset-ting plastics passed; a decade in which they demonstrated their possibilities for the speedy and economical production of high quality products. The applications of the period show how they reduced assembly operations and promoted production economy. Tube bases, (fig. 5–19) previously assembled in 10 operations, were molded complete in 1 operation, and the cost was reduced 36%; machine parts requiring as many as 9 operations,

Figure 5–21   International Radio Company pioneered the first plastics radio cabinet in the United States in 1931.

were molded in only 2, eliminating turning, boring, threading, and milling.

In 1930, twenty million people, listening to the famous *Amos 'n' Andy* program, heard an announcement covering the use of Bakelite caps for a new Pepsodent bottle. This was plastics promotion at its best, taking advantage of the tremendous popularity of the heartwarming and mirth-provoking dilemmas of Amos and Andy. The

Elgin watch box (fig. 5–20) designed by Ted Sloan and molded by Chicago Molded Products Company, initiated a trend to packages which also displayed the merchandise nicely.

In 1931 Ultra Midget radio (fig. 5–21) used a phenolic case that was pioneered by International Radio Company. This was followed by their famous Kadette AC/DC radio in 1932 and started the big boom in radio cabinets. Band-Aids were first

Figure 5–22   These leading industrial designers were brought together in 1932 for a plastics symposium conducted by Bakelite Corporation to expand their knowledge of the potential uses for plastics. Shown here are Helen Dryden, Gustave Jensen, Norman Bel Geddes, Lurelle Guild, John Vassos, Joseph Sinel, Lucien Bernhard, Donald Deskey, George Switzer, Simon de Vaulchien, Henry Dreyfus, George Sakier, and Walter Dorwin Teague.

waterproofed with phenolic resin in 1931, and a phenolic ash tray was offered then as a sales premium.

The Great Depression years were times of innovation and giant inroads for thermosetting plastics. Thermoplastics were beginning to appear, but they were often ignored by the thermosetting molders. Manufacturers were forced to offer new values, their life depended on redesigned products that cut costs, improved appearance, simplified construction, and offered more plus values.

A most effective promotion for plastics was made by Bakelite Corporation in 1932, taking advantage of growing interest in the trend toward modern design and the intention of manufacturers to gain sales by improved appearance and tactile properties. Bakelite held symposiums on design and conducted forums and conferences with leading industrial designers. Norman

Figure 5–23 The world's first electric shaver, Schick Model A, was molded of red phenolic by Norton Laboratories in 1930. The mold cost only $400.00 at that time.

Bel Geddes, Walter Dorwin Teague, Joseph Sinel, Lurelle Guild, John Vassos, Lucien Bernhard, Donald Deskey, Henry Dreyfus, Simon de Vaulchien, George Switzer, and George Sakier all participated in the initial effort (fig. 5–22). A continuing program of education by Bakelite, Durez, and Plaskon reached all active industrial designers, who contributed greatly in the swing to plastics. One manufacturer of home instruments redesigned his product attractively with a phenolic housing. In 1931 his sales were 43% above 1930 and 90% above 1929. In 1932 his sales broke all records. This was at the bottom of the Great Depression.

The first electric shaver (fig. 5–23), produced early in 1930, greatly accelerated the interest in plastics and opened several other markets. Schick's Model A was molded by Norton Laboratories. The Model B was molded of black phenolic by

Figure 5-24 *The second Schick electric shaver used the famous bridge construction for its motor and was molded in 1931 by Shaw Insulator Company.*

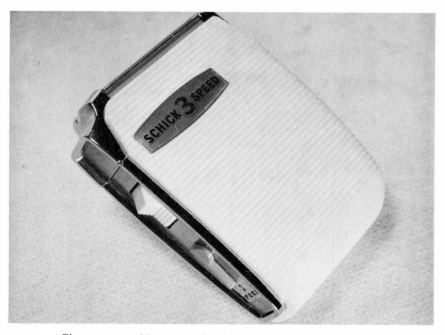

Figure 5-25 *Many of the pioneering applications that started with thermosetting plastics were converted to thermoplastics after World War II. The Schick shaver was molded of ivory urea for many years. It was converted in 1959 to nylon by Foster Grant Company. Model 3 in nylon is shown here. (Courtesy, Foster Grant Company.)*

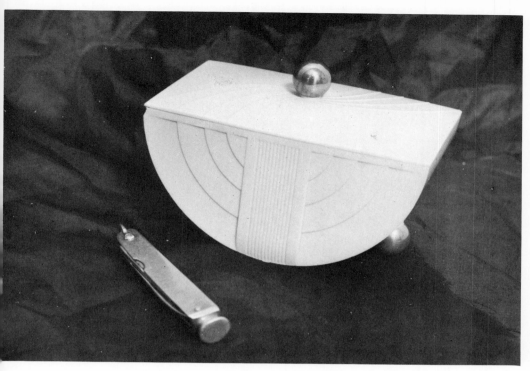

Figure 5–26   The Kool cigarette box was used as a self-liquidating premium to introduce the menthol cigarettes. Designed by Ted Sloan and timed with the world's fair, A Century of Progress, this outstanding package established a new high in premium promotional practice.

Shaw Insulator in the fall of 1931 and used the unique bracket construction (fig. 5–24) for the motor, which took advantage of a very complex insert assembly and advanced molding techniques. Urea materials were used by Schick for many years prior to its conversion to nylon in 1959 by Foster Grant Company (fig. 5–25). Plaskon retained Van Doren and Rideout to expand their markets via better design; Chicago's Jean Reinecke spe-

cialized in designing with plastics and improved many products. General Electric's Bill Petzold advanced the use of phenolics and ureas in business machines, lighting fixtures, appliances, and cosmetic packages. The Wheaties and Skippy bowls and the the Orphan Annie mug, all made of urea, achieved unprecedented sales records and started a decade of merchandising with plastics premiums.

The plastics premiums of the

Figure 5–27   This exhibit of plastics products produced by General Electric Company is typical of many displays a the 1933 world's fair, A Century of Progress.

1930s were very important salesmen for the plastics industry: they got samples of these new and unknown products into the home. Gage Davis, who was then advertising promotion manager at General Mills, reports the following on their most successful use of plastics:

The Skippy bowl was a very popular colorful urea plastics premium. The comic strip character, "Skippy" was molded in relief in the bowl. 5,000,000 pieces of this popular premium were used by General Mills, giving one free with two packages of Wheaties. Plastics was a novelty and part of the appeal was its newness and the crisp lines not achieved with the ceramics or glass. 12,000,000 Bisquick cookie cutters were distributed by General Mills in the 1932 and 1933 era. Based on 35,000,000 homes in that period, this put plastics into one of every four homes in the United States and created real interest in the plastics.

Ted Sloan's 1933 Kool cigarette box (fig. 5–26), symbolizing a Cen-

Figure 5–28   Steel mill roll-neck bearings were converted from bronze and babbit to laminated phenolic in the early 1930s. Power savings as high as 40% were reported by many mills. (Courtesy, General Electric Company.)

tury of Progress and introduced at the Chicago Fair, broke all records in cigarette promotional plans. It introduced Kool cigarettes on a self-liquidating basis. Plastics premiums, packaging, housings, toys, and novelties became a big tool to fight the depression.

Numerous displays at the Chicago World's Fair, such as that shown in figure 5–27, demonstrated plastics progress and expanded interest at all levels. Many other ex-

hibitors such as Schick, Formica, Westinghouse, Bakelite, and others featured plastics products. Of special interest at that time were the Bakelite clear phenolic labware and watch crystals, Marshalltown Trowel Company's break-resistant handles, Koken and Textolite golf clubs, Electric City's standardized line of jewelry cases.

Phenolic laminated roll-neck bearings for the steel mills, introduced in 1935, established new

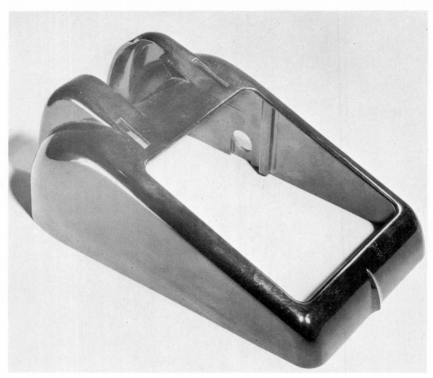

Figure 5–29   This 1938 housing for the Victor Adding Machine Company was the pioneer application in the field of the business machines. Durez promoted work in this field. (Courtesy, Victor Adding Machine Company, Chicago, Illinois.)

speeds and economy in steelmaking procedures. The power savings introduced by plastics (fig. 5–28) compared with babbitt-bronze bearings were so phenomenal that it caused one major steel company to have a study made of the potential effect of plastics on the metal industry's future. That 1935 report envisioned none of of the real competition that came twenty-five years later. Piano keys, the ebonies and the ivories, made use of plastics.

Sharps went from ebony to thermoformed Celluloid to phenolic. Hammond's 1935 electric organ utilized urea for natural keys, replacing the Celluloid that had superseded ivory. Urea was later supplanted by injection-molded styrene acrylonitrile, and now a comeback is being made by injection-molded melamine to gain the hardness and abuse resistance of thermosets.

The pace-setting Toledo Scale housing (fig. 5–3) of 1935 triggered

Figure 5–30   This cabinet for the Pilot all-wave radio was one of the first large phenolic cabinets. It was molded in 1936.

a deluge of machine housings such as the Hobart scale and grinder, the Victor adding machine (fig. 5–29), the Stenotype machine and the large Pilot all-wave radio cabinet (fig. 5–30). The 36-inch Wakefield lighting bowl and the Galloway cream separator bowl (fig. 5–31) were typical plastics items of progress in the late 1930s. Some 1935 products are shown in figure 5–32.

Colorability of the cast phenolics gave them a brief position in the 1930s. Ortho Plastics Novelties, headed by Oscar Gold, became a large fabricator of Catalin and Marblette. In 1936 they fabricated a cast phenolic radio cabinet (fig. 5–33) that was distributed by Monsanto to RCA, Emerson, and FADA. Cast phenolic resins were produced by Bakelite Corporation, Catalin Corporation, A. Knoedler Company, Marblette Corporation and Monsanto Chemical Company.

The Bakelite Travelcade, initi-

Figure 5–31 The resistance of the phenolic plastics to the solutions used to clean milk-handling equipment was recognized in the production of this bowl for a cream separator. (Courtesy, General Electric Company.)

Figure 5–32 Shown here is a diversified group of plastics products as produced in 1935. Noteworthy is the absence of many thermoplastics. (Courtesy, General Electric Company.)

Figure 5–33 Cast phenolic radio cabinets had a brief period because of their colorability and marbleized effects. Shown here is a finishing operation. (Courtesy, Ortho Plastics Novelties.)

ated in March 1938, took plastics down Main Street of America bringing to the public the real story of modern plastics. Its theme was, "What Is This Plastics Business?" Animated exhibits, lectures, symposiums, sound pictures, and dramatic presentations were made to describe the genesis and workings of the plastics industry. A sound film, "The Fourth Kingdom," narrated by Lowell Thomas, described how research has taken the three kingdoms, vegetable, animal, and mineral and created a fourth—plastics. Typical major exhibits included packaging, display, fashion, radio, industrial components, and housewares.

The Travelcade was viewed by thousands of people and ended its trip at the Franklin Institute in Philadelphia with a clinic on bonds for plywood.

The latter part of the 1930s saw labor rates increasing to the point

Figure 5–34 This 1933 iron handle of phenolic molded by Chicago Molded Products for Birtman Electric Company pioneered the one-piece iron handle of plastics.

where it became uneconomical to use cheaper raw materials when the part could be molded or stamped in final form from the plastics. Wood was cheap but, with high labor costs, paintbrush and tool handles went to plastics with molded-in finishes and with high resistance to solvents and household chemicals. In 1933 Birtman Electric Company pioneered the electric-iron handle shown in figure 5–34, styled by Alphonse Ianelli. This started the conversion of all iron handles to plastics.

In the late 1930s war was in sight, metals were being restricted to "defense" uses, brass and copper were foreseen to be in short supply. Many products such as the milking-machine parts shown in figure 5–35 were converted to plastics.

Giant molding plants with highly mechanized equipment had been developed as shown in figure 5–36. The development of aircraft com-

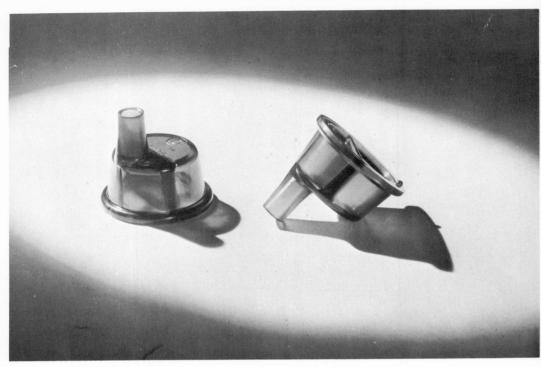

Figure 5–35   These milking-machine parts were molded of clear Bake-lite resin in anticipation of restrictions on the use of brass for non-priority items during World War II.

Figure 5–36   This view of a 1938 General Electric Company molding plant is typical of the facilities before World War II. (Courtesy, General Electric Company.)

Figure 5–37 The frangible bullet was transfer molded of a lead-filled phenolic compound in 1945. This bullet gave the gunners live target practice on a friendly plane without hazard or damage. The frangible bullet splattered when it hit the target plane.

ponents by the laminators was starting, and competition from the thermoplastics was in the air.

World War II

Edward R. Stettinius, Jr., Priorities Director during World War II, focused the eyes of industry on plastics on March 16, 1941, when he advocated the substitution of plastics for aluminum, brass, and other strategic metals. The thermosets became all-powerful in World War II. Critical metal shortages, mass production of intricate components, radar direction finders, plastics-bonded wood or fabric for aircraft components, gliders, and boats all exploited the thermosets and their fabrication facilities. Tremendous molding plants grew up all over the country and know-how was widely disseminated. The frangible bullet (fig. 5–37) and the M-52 trench

Figure 5–38   The first million-dollar order for plastics was for the M-52 trench mortar fuze. Many molders produced these parts in very high volume.

mortar fuze made plastics history in volume, in quality control, and in field performance. Components of the fuze are depicted in figure 5–38. The loop antenna housing shown in figure 5–39 was molded of a cloth-filled phenolic, and it appeared on all of the war aircraft. Sound power telephones, instruments of all types, and nameplates (fig. 5–40) were made of plastics. Plastics were used to lighten aircraft components and to replace brass and aluminum in all nonconducting applications. Bomb-burster tubes and bazooka barrels were made from the laminates. Navy dinnerware went to melamine. Plastics for civilian items were not easily available.

This was a period of great expansion, experiment, innovation, and development, generated at an extravagant place and with tremendous enthusiasm. The thermoset-

Figure 5–39   The loop-antenna housing was the radome for aircraft guidance. It was molded of fabric-filled phenolic in two halves that were bolted together. (Courtesy, General Electric Company.)

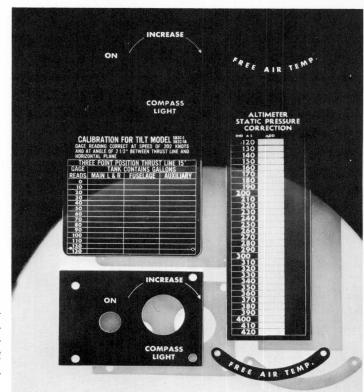

Figure 5–40   Laminated plastics nameplates were used for all classes of applications during the war. They were noncorrodible, highly visible, and easily made.

Figure 5–41 Polyester glass compounds solved many of the circuit-breaker problems of war and peace. (Courtesy, Haysite Division, Synthane Taylor Corporation.)

ting materials hit their peak of performance and acceptance during the war.

Alkyd materials (fig. 5–41) and polyester resins were introduced during the war, and their low-pressure moldability indicated their use for encapsulation. They surpassed the performance of many older thermosets for low-loss electrical applications and for certain electronic compounds. Polyesters paved the way for the low-pressure lami-nates now known as reinforced plastics (covered in chapter 9).

When Henry Ford swung an axe (fig. 5–42) on one of his experimental plastics automobile bodies in 1945, he dramatized the trend and focused worldwide attention on the plastics materials as potential competitors of steel.

Dr. Howard L. Bender, Assistant Director of Research and Development, Bakelite Corporation, received the Hyatt Award in 1953 for

Figure 5–42   Henry Ford used an axe to demonstrate the potential value of the plastics for automobile bodies and other structural shapes. This 1945 demonstration alerted the world to the postwar plastics applications. (Courtesy, Ford Motor Company.)

his lifetime of research and development in the field of phenolic resin molecular structures.

The Glastic Corporation concentrated on the polyester and alkyd glass products and quickly took over the brush stud market (fig. 5–43). Their Glastic sheets filled many markets in the circuit-breaker field that had been served previously by other laminates, the ceramics, ceramoplastics, and cold-mold products.

The allyl plastics became available in 1955 and were produced by FMC. Mesa Plastics Company, founded by Felix Karas, pioneered these important materials for electronic and electrical applications. The tiny insulators shown in figure 5–44 are used in very precise and highly technical applications. These materials filled an important gap in the missile program and gained favor because of their excellent chemical resistance, low electrical

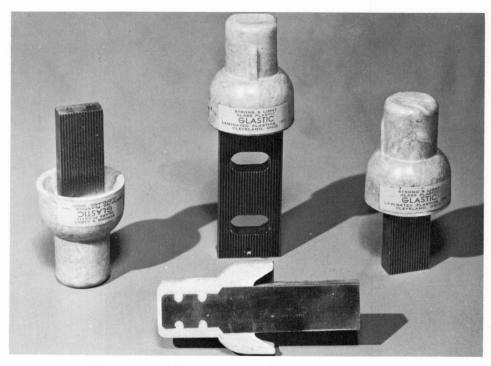

Figure 5–43 *These polyester glass brush studs of 1948 offered adequate strength, arc resistance, and dielectric strength to serve in railway and mine locomotive applications. (Courtesy, Reliance Electric Company.)*

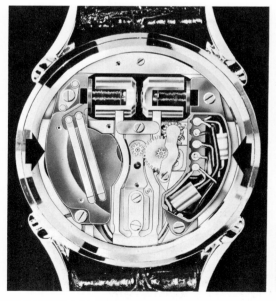

Figure 5–44 *Two parts molded of a diallyl phthalate resin-based compound serve as the chassis for the complete coil assembly (arrows in left photo) of the Accutron electronic timepiece. Magnification of the unassembled chassis (at right) reveals the critical tolerances that the parts must meet.*

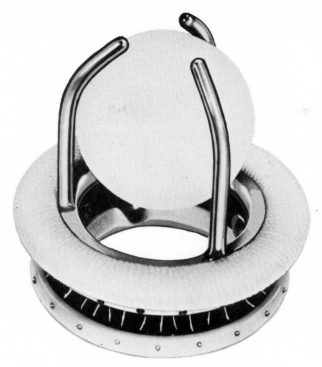

Figure 5–45 The Starr-Edwards heart valve shown here has a chrome-cobalt cage, Teflon sewing ring, and a silicone ball. This was developed by Dr. A. Starr and Lowell Edwards.

loss, weatherability, minimum shrinkage, inertness, and stability without galvanic corrosion.

The silicones are a chemical hybrid—in between the organic plastics and sand. Aside from being one of the best high-temperature organic plastics, their markets include rubbers, water repellents, lubes, polishes, coatings, mold material, adhesives, potting compounds, sealants, high-temperature fluids, and lubricants. Silicone rubber products are extensively used for heart-valve components (fig. 5–45), prosthetic devices, and reconstructive surgery. Silicones were anticipated by an Englishman, Frederick S. Kipping, early in the twentieth century, but he did not envision any commercial potential for them. Corning Glass Company's J. F. Hyde made polymers of silicone and oxygen in 1931 and gained additional help from Dow Chemical Company. They formed Dow Corning Corporation in 1943 to produce silicones by a process suggested by Kipping. Corning's demonstration to General Electric Company aroused further interest and Drs. W. I. Patnode and E. G. Rochow were assigned to do advanced development work. General Electric Company also went into production in 1943 using Rochow's process. These products were followed in 1965 by the Owens Illinois glass resins.

Silicone rubber has been used for a variety of heart valves over the years. It is believed that the first mention of this use of silicone rubber was made by Dr. C. P. Bailey in the late 1940s. The first implantable Pacemaker was developed by Dr. William H. Chardack in the late 1950s. The Pacemaker is encapsulated in silicone rubber; the electrodes are covered with heat-vulcanizing silicone rubber. The silicones have been used for rubber ear armatures, rhinoplasty implants, with Dacron felt for chin implants, for mammary prosthesis and numerous other human body parts. The Akutsu-Kolff artificial heart is made of polyester-reinforced silicone rubber.

The epoxy resins were invented by a Swiss, Dr. P. Castan, in 1938. Shell Chemical Corporation introduced them to the United States, working with Devoe and Reynolds in 1941 for surface coatings. Drs. E. Preisweek and A. Gans discovered the tremendous properties and unique capacity to unite a wide range of materials such as metal and glass. Today some automobile body components are "welded" with epoxy, and epoxy molding compounds are creating multiple new markets for thermosets.

The Era after World War II

New tooling for the pent-up de-

Figure 5–46   This first television cabinet was molded in 1949 of walnut and mahogany color phenolics. It weighs 35 lb and was used by Admiral Corporation.

mand for appliances, automobiles, housing, dinnerware, housewares, and toys in the postwar era took advantage of the wartime methods and facilities. There was no loss of work in plastics during the conversion from war to peace. The 1949 Admiral 35-lb. TV cabinet set another milestone in large plastics parts (fig. 5–46). The thermoplastics clouds were just beginning to hang over the molders of thermosets; the revolution had begun. Radio cabi-nets were changed overnight from urea and phenolic to styrene and acetate compounds, gaining toughness and colorability. In a single season, an entire large market for thermosets was lost to the thermoplastics. Pressure on the underwriters gained approval for other household and electrical devices that used thermoplastics with self-extinguishing additives. Nylon did many jobs better than the cheaper thermosets and often reduced the cost because

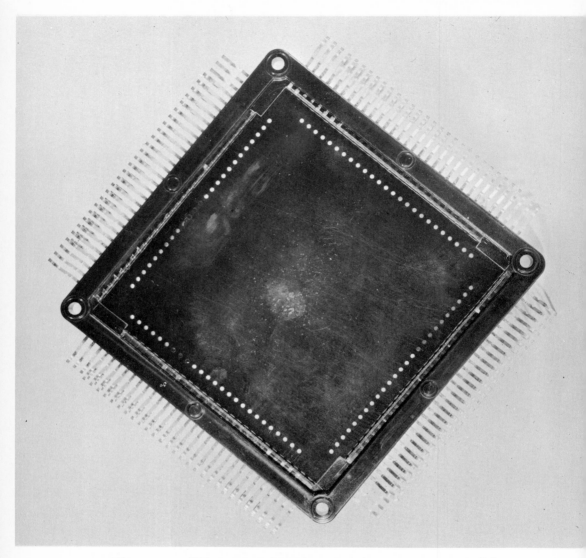

Figure 5–47 This memory-plane component has two rows of parallel contacts. It was used in the Univac computer. The requirements of the computer industry demanded the best of the know-how and materials that had been developed during World War II. This piece was transfer molded by Tech Art Plastics Company in the middle of the 1950s. (Courtesy, Tech Art Plastics Company.)

Figure 5–48 This Stokes Injectoset machine introduced in 1963 triggered a resurgence of thermosets enabling the thermosetting molders to recover many jobs that had been lost to the thermoplastics. The machine shown here is molding keys for piano sharps on a fully automatic cycle with speeds comparable to those of the injection molding of thermoplastics. (Courtesy, Tech Art Plastics Company.)

of faster cycles and minimum mold requirements.

Fortunately, the expanding industrial growth of the postwar economy and the upswing of the missile and computer age with the demands for extreme sophistication created other new markets for thermosets, so that they continued to grow, but at a reduced rate. A typical computer memory center is shown in figure 5–47. Early in the 1960s, marketing of the Stokes Injectoset (fig. 5–48) focused attention on a revived potential for the thermosets to regain some of their lost markets by using the same high-speed molding procedures (chapter 6) that had built up the thermoplastic takeover. Special compounds were developed by phenolic resin makers to facilitate the use of these resins in the automatic molding machines.

Advanced electronic circuitry design pressed the need for improved

Figure 5–49   This special encapsulation press was designed by Hull Corporation for the encapsulation of electronic components. Special mold frames are provided to facilitate the loading of many inserts in each cycle.

Figure 5–50   The polyester glass materials of World War II moved into everything from the Sousaphone to an integral bathroom. This Sousaphone weighs 43% less than the all-metal model and has a larger and richer tone. (Courtesy, Conn Corporation.)

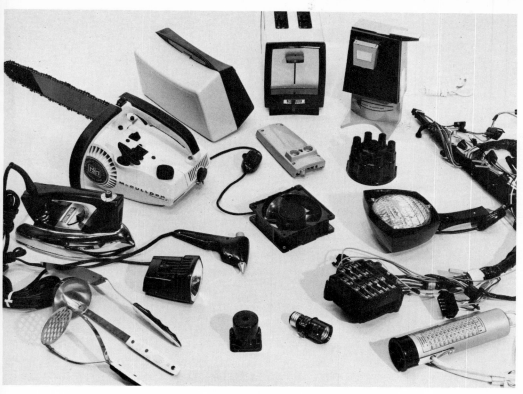

Figure 5-51   Newly developed phenolic compounds suitable for fully automatic molding have lowered the cost of thermosetting products. (Courtesy, General Electric Company.)

encapsulation methods and materials. Potting with cast resins initiated this procedure at National Bureau of Standards during World War II. Transfer molding offered an excellent procedure for high-volume production. Hull Corporation made special studies of this market and produced special encapsulation presses (fig. 5–49) that are widely used by electronic component makers. Durez and General Electric en-

gineers developed special high-flow, high-durability phenolic compounds to compete with the epoxy materials in this market.

The molded fiber-glass materials (chapter 9) permitted use in many unique applications (fig. 5–50), including automobile bodies.

Many products that really need better rigidity, surface hardness, and self-extinguishing properties moved back to thermosets (fig. 5–51). It is a healthy struggle, bringing a continuing challenge to makers of materials and machines and to all molders and fabricators. Custom molders today face the same challenge that their forbears faced in 1915. They develop new applications and solve production problems for their customers. When the volume builds up, the customers put in their own molding machines, and the custom molder looks for new customers.

6

Developments in Thermoplastics Machinery and Method

Many of the thermosetting molders used conventional compression molds on a heat-and-chill basis for the pioneer molding of acetates, as told in chapter 7. This procedure was suggested by the materials makers before injection machines were recognized and available. Even then, there was considerable fear of change and prejudice against the new methods. One large molding plant superintendent was heard to say, "We have enough problems with the things we know how to do without tackling something we don't understand." The result in that plant was one Isoma injection machine standing in its crate for a year before it was tried out. In the interim, cellulose acetate Ford knobs were being compression molded on a 5-minute heat-and-chill cycle. The mold was first heated with steam and then cooled. Cooling was achieved by circulating water from a horse trough containing two refrigeration units.

The injection machine first came to the United States in 1922, but it took 10 years to gain adequate recognition and serious general interest; material problems were partially responsible for the slow development and acceptance.

The year 1936 may be said to be an awakening year for the adolesent injection molding process. Cellulose acetate, shellac, and polystyrene were being marketed and the basic facts had been published. It is significant that the *Modern Plastics Encyclopedia* for 1936 contained advertisements for only 5 injection machines, Burroughs, Index, Southwark, Krehbiel, and Hydraulic Press Manufacturing Company (HPM). The machines were then called extrusion molding presses. No extruders were advertised for sale in 1936. In 1938, they were called injection machines and were

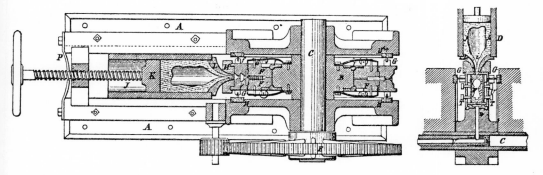

Figure 6–1 In 1870 John J. C. Smith and Jesse A. Locke patented in the United States this mechanical device for "the manufacture of castings under pressure." It is the prototype of the ram-injection machine. They used it for the die casting of metals. (Courtesy, "The Historical Development of Plastics Processing Techniques," Kunststoffe, vol. 55, March 1955.)

offered by HPM, DeMattia, Lester, Reed-Prentice, Watson Stillman, and Grotelite. In 1970 more than 50 manufacturers in the United States sold injection machines.

The thermoplastics injection machine* was invented four times. John C. Smith and Jesse A. Locke in 1870 invented the machine illustrated in figure 6–1 for the "Manufacture of Castings Under Pressure." Their machine is a real prototype of the conventional plunger injection molder which preceded the screw injection machines. It was used for die casting of metals, which is very similar to injection molding. Their pressure chamber had a piston driven by a spindle. The injection cylinder was exchangeable, and their cylinder was closed by a removable nozzle designed to form a "tear-off" gate.

The Hyatt Brothers machine

* These details were recorded in *Kunststoffe*, vol. 55, March 1965.

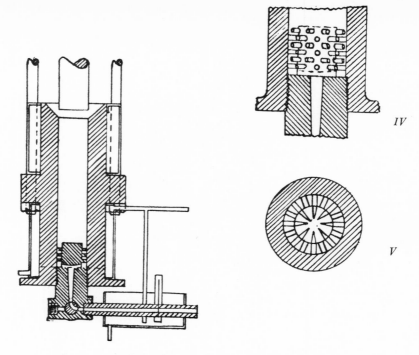

IV

V

Figure 6–2   Shown here is the patent drawing for the original Hyatt injection-molding machine, patented in 1872. Noteworthy are the heat-transfer points illustrated in views IV and V, which brought the heat into more intimate contact with the material and anticipated the Gastrow torpedo.

(fig. 6–2), which is also reported in chapter 2, was invented in 1872. Hyatt fed Celluloid preforms into the heating cylinder and then pushed the plasticized charge into the mold with a piston. Hydraulic pressure was used for the piston extruder. Three of these machines ran for many years, and this injection molder was used as evidence of prior art in a patent suit 70 years later with Bucholz and Eichengrün.

The Englishman E. L. Gaylord patented an injection process in 1904 (fig. 6–3). He used it for molding amber, which is similar to styrene, but the high material cost limited his market.

A. Eichengrün succeeded in making an injectable cellulose acetate material in 1919 and invented a machine that was built by H. Bucholz. The Eichengrün-Bucholz machine's schematic design is shown in figure 6–4. This was the fundamental prototype of all the early

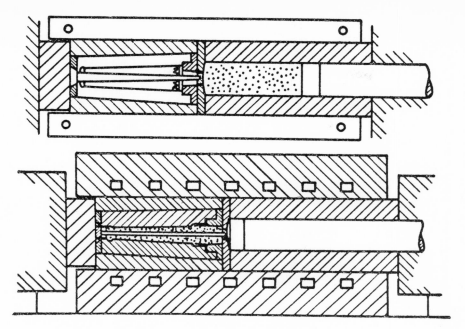

Figure 6–3   This 1904 injection machine was developed in England by E. L. Gaylord for the injection molding of amber.

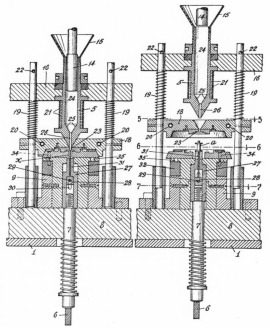

Figure 6–4   This injection machine was designed by H. Bucholz following A. Eichengrün's making of an injectable cellulose acetate in 1919. It is the precursor of the early American injection machines. Radio Corporation of America built machines of this type for its own use during World War II.

Figure 6–5 Mr. W. D. Grote of the Grotelite Company imported 12 of these Bucholz machines in 1922 and thus became the first American injection molder of plastics. (Courtesy, Grote Manufacturing Company, Madison, Indiana.)

injection machines in the United States after Hyatt.

Bucholz filed an American patent application in 1923, which was granted 8 years later in 1931. This patent was assigned to W. D. Grote of the Grotelite Company, Bellevue, Kentucky (now Grote Manufacturing Company of Madison, Indiana). Mr. Grote had imported 12 of these machines in 1922. They were the first injection machines after Hyatt to produce molded parts in the United States. One is shown in figure 6–5. Celanese reported their injection molding work on an Eckert and Ziegler machine in 1927. The machine was fed with a teaspoon and gas heated. Celanese had purchased American Salon Company and obtained 2 small Eckert and Ziegler machines.

From the original hand-operated Bucholz machines, the Grotelite Company developed the first fully automatic injection molding press

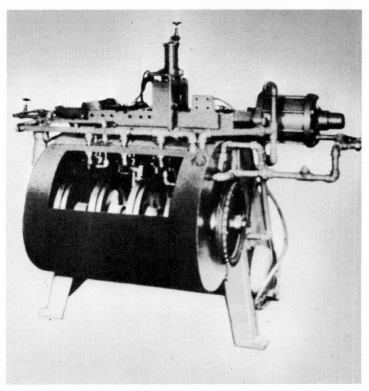

Figure 6–6   This is the first fully automatic injection machine, built by Grote Manufacturing Company in 1929. (Courtesy, Grote Manufacturing Company, Madison, Indiana.)

in 1929 (fig. 6–6). This was followed by their development of the first hydraulically operated automatic press in 1933 (fig. 6–7). In 1933 four of these latter machines were sold to Du Pont Viscoloid, which put Du Pont into the injection molding business. Du Pont also received drawings of these machines and had a quantity built by a New England manufacturer. Shown in figure 6–8 is the first Foster Grant injection molding machine. It was built by Eckert and Ziegler and purchased by Samuel Foster in 1930. Foster Grant built 20 such machines in the next 3 years; they added the Gastrow torpedo spreader in 1932 and built their first hydraulic machine in 1934. Shown in figure 6–9 is one of the early commercial machines as sold by HPM in 1931. Many of the American machines were home made and hand powered. Van Dorn introduced and sold many such ma-

Figure 6–7   Grote Manufacturing Company built this first hydraulically operated automatic injection machine in 1933. (Courtesy, Grote Manufacturing Company, Madison, Indiana.)

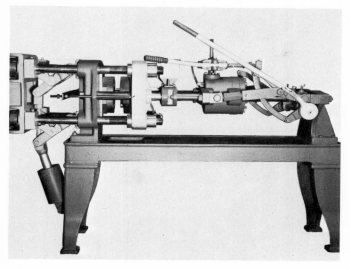

Figure 6–8   This 1930 Eckert and Ziegler injection machine was imported from Germany in 1930 by Samuel Foster of Foster Grant Company.

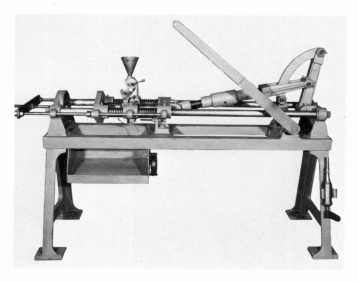

Figure 6–9   This 1931 HPM injection machine is typical of many of the early machines. Many were built by the molders. (Courtesy, Hydraulic Press Manufacturing Company.)

Figure 6–10   This hand-powered machine, based on the Bucholz design, was first marketed by Van Dorn in 1946 and points up the stability of that original design.

chines (fig. 6–10) in 1946. During World War II, RCA and others built their own machines, that were close followers of the old Bucholz design.

The Hyatt patent for the 1872 injection molding machine shows a central core (fig. 6–2, IV and V) or restrictive area similar to a torpedo that forced plastics through the thin open space to achieve maximum thermal transfer. Many pegs will be seen between the cylinder wall and the restrictive core, added for the transfer of heat by conduction.

The transfer of heat in adequate quantity to fully plasticize the material limited the size of early injection machines. In 1932, the invention of the Hans Gastrow torpedo (fig. 6–11) increased greatly the plasticizing capability, facilitating the construction of larger-capacity machines with a single cylinder.

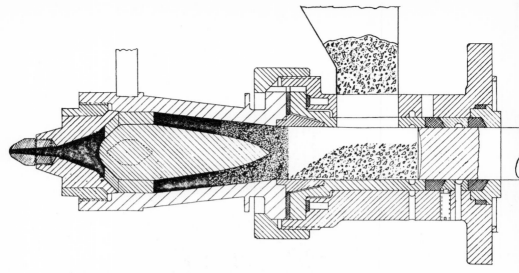

Figure 6–11 *The invention of the torpedo in 1932 by Hans Gastrow increased the plasticizing capacity of the injection machine.*

Figure 6–12 *This power-driven injection machine was built by E. W. Bliss in 1934 for Tennessee Eastman Company and pointed the way to the larger machines.*

Figure 6–13   This early Isoma injection machine, sold by Index Machinery Company, was advertised in 1936 as the first clock-controlled injection machine.

This streamlined section increased the thermal contact area for greater heat transfer.

In 1934, Tennessee Eastman Company challenged the validity in the United States of the Bucholz patents that had been acquired by Celanese from American Salon Company. Tennessee Eastman notified all molders that they would defend any infringement suit. The Bucholz patents were declared void in 1942 based on the prior art as disclosed by Hyatt in 1872.

E. W. Bliss built a power-driven unit for Tennessee Eastman Company in 1934 (fig. 6–12) which was a precursor of the power machines of later years. A fully automatic, clock-controlled injection molding machine as shown in figure 6–13 was sold by Index Machinery Corporation in 1936 and built by Isoma. Many Isoma machines were

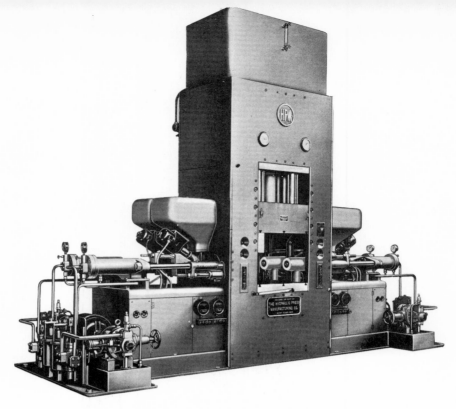

Figure 6–14   Hydraulic Press Manufacturing Company built this 36-oz machine in 1936 by the use of four 9-oz injection cylinders.

imported by American molders. Manufacturers of compression presses and die casting machines jumped into the act and several types soon became available.

The next step was the use of multiple injection cylinders. Figure 6–14 shows a 36-oz press made in 1936 by Hydraulic Press Manufacturing Company. It used four 9-oz injection units. The quadruple gating therefrom offered some mold design problems to ensure good welds at the flow juncture points. Larger capacity cylinders and the two-stage units offered the means to much larger machines up until the advent of the reciprocating-screw plasticizing injection unit in 1957.

The first Reed-Prentice injection machine was built in 1936 as shown in figure 6–15. The 10-D-8 machine (fig. 6–16) built in 1938 by Reed-Prentice made all-time history. It became the world's most popular injection machine; over 2,000 were

Figure 6–15 *Reed-Prentice built its first injection-molding machine in 1936. This is a 2-oz machine. (Courtesy, Package Machinery Company, East Long-meadow, Massachusetts.)*

Figure 6–16 *This Reed-Prentice 10-D-8 made injection history after its introduction in 1938. It was then the most popular injection machine in the world. Over 2,000 were built, and many are still running. Some have been converted to screw feed. (Courtesy, Package Machinery Company, East Long-meadow, Massachusetts.)*

Figure 6–17   This early HPM injection machine used a hydraulic clamp. (Courtesy, Hydraulic Press Manufacturing Company.)

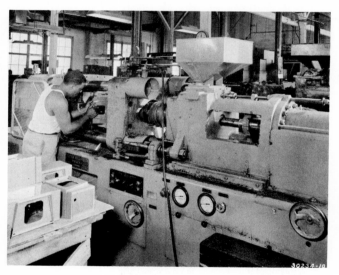

Figure 6–18   This early Watson-Stillman injection molding machine featured a toggle clamp. Here it is shown molding some of the first thermoplastic radio cabinets.

Figure 6–19   This battery of injection-molding machines at Foster Grant Company is typical of the larger injection plants of the early 1950s.

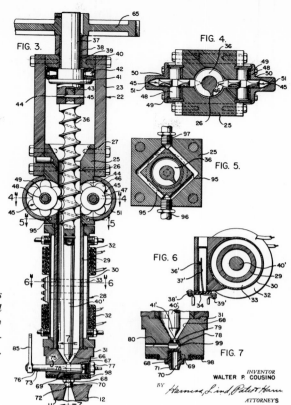

Figure 6–20   The Cousino-Chrysler press of 1943 was an early attempt to mold thermosets and thermoplastics on fully automatic cycle. It was used more successfully for rubber products.

built, and many are still running —32 years later. Other popular machines of the forties are shown in figures 6–17, 6–18 and 6–19.

In 1943, W. P. Cousino developed an injection machine (fig. 6–20) which made use of a self-centering spreader that rotated to bring the material into contact with the heater wall. This was also conceived to minimize excessive temperature build-up between cycles by a reduction of pressure when

thermosets were used. Chrysler used this press very effectively for the molding of rubber compounds.

Improved Machinery Incorporated entered the injection molding machinery field in 1945 with a vertical-clamp, auxiliary-compression, horizontal injection unit as shown in figure 6–21. This machine was particularly successful in the molding of void-free heavy-section molded parts. In operation, the mold was clamped in the vertical

Figure 6–21 This 1944 press of the Improved Paper Machinery Corporation was made first to solve the heavy-section acrylic parts. It was used to develop the acrylic feeder heads used to produce potable water in the Pacific Islands by the armed forces. Compression-molded parts needed a 1-hr cycle to make the heavy section. This press permitted a combination of injection and compression molding without loss of high pressure during the cooling cycle. (Courtesy, Improved Paper Machinery Corporation, Nashua, New Hampshire.)

position, and injection came in from the side, filling the mold. At this point the mold force pushed up into the cavity, closed the gate, and densified the cooling thermoplastic material. In this manner, continuous pressure followed the contraction of the cooling plastics part and eliminated sinks and distortion. This injection-compression press has been very successful in the molding of lenses and phonograph records.

Screw preplasticization for the thermoplastics was proposed in 1932, but nothing of record was done. Jim Hendry built a 2-oz screw thermoplastic injection machine in 1946. This was followed in 1948 by his two-stage machine, using a screw plasticizer and piston injection, which ran in Saginaw, Michigan. Jackson and Church then built a 48-oz two-stage machine followed by a 64-oz unit. As shown in figure 6–22, these were the first machines that

Figure 6–22   This is the first two-stage machine as built by Jackson and Church in 1948. It was the first successful machine for the molding of un-plasticized PVC and represented a great gain in the art.

Figure 6–23   During the transition of many jobs from thermosetting materials to thermoplastics, many molders added injection cylinders to their old compression presses to move into the injection field. These presses were fitted with injection cylinders by Eric Gronemeyer and Don Kendall at Mack Molding Company. (Courtesy, Mack Molding Company, Arlington, Vermont.)

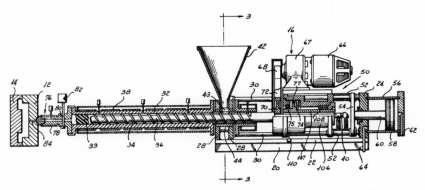

Figure 6–24   This invention, patented in 1956 by W. H. Willert, revolutionized injection blow molding and the injection of thermo-plastics and thermosets.

could mold unplasticized PVC and were a major milestone in thermo-plastic molding.

During the late 1930s and the following war years, many compression presses were made into injection machines. Side rams were sometimes added for the injection system; others mounted a top ram for injection following the practice of the compression presses converted to transfer. Figure 6–23 shows a battery of converted compression presses in the Mack Molding plant in Arlington, Vermont.

The invention by William H. Willert of the in-line or reciprocating screw plasticizing injection unit (shown in figure 6–24, filed in 1952 and issued in 1956) brought the first major improvement to injection machines since the addition of the torpedo.

The action of an extrusion screw in a temperature-controlled heating cylinder provides better and more

Figure 6–25 Reed-Prentice built this 600-ton reciprocating-screw press in 1953, the first to use Willert's invention.

uniform mixing of all compounds in comparison with the plunger-plasticizing injection system. Melts are more uniform because of the homogenizing action of the screw and the reduced overall temperature. The reciprocating screw, which also performs as a plunger, has been found to be the most advantageous. Taking advantage of a nonreturn valve on the end of the screw, which controls back pressure on the screw during the plasticizing period and limiting the back travel of the screw, facilitates metering the exact volume of material needed to fill the mold. No pressure is lost; drool and waste are eliminated. Another gain results from the removal of gas as the material moves along the screw.

Shown in figure 6–25 is the first reciprocating screw injection press as built by Reed-Prentice in 1953. Egan Machinery Company built an experimental reciprocating screw

Figure 6–26 Egan Machinery Company built their first Reciproscrew and mounted it on this HPM machine for Du Pont in 1958. The resulting Du Pont studies popularized the screw injection machine and revolutionized injection molding. (Courtesy, Frank W. Egan & Company, Somerville, New Jersey.)

plasticizer and assembled it on an HPM clamp (as shown in figure 6–26) for Du Pont in 1958. Bitter debate and many erroneous statements followed the introduction of the first commercial reciprocating screw machines, but the screw plasticizer won over the piston type by its superior performance and is used almost exclusively today.

The late 1950s saw the development of completely automated injection molding plants pioneered by Eastman Kodak Company. In these plants all operations were under a master control and were conveyorized to eliminate handling. Scrap was ground and blended automatically with precise control. Shown in figure 6–27 is a highly automated Ford Motor Company injection plant of the late 1960s.

In the early days of injection molding, cold tap water was used for mold temperature regulation. The advent of higher-heat thermo-

Figure 6–27 Shown here is a 1969 high-volume injection plant at the Ford Motor Company.

Figure 6–28 Mold chillers such as this Application Engineering Corporation's 40-to-105-ton model will maintain low-mold temperatures for thermoplastic molding.

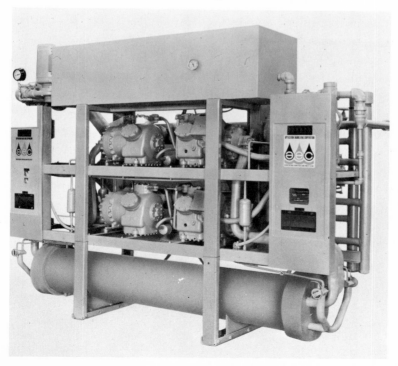

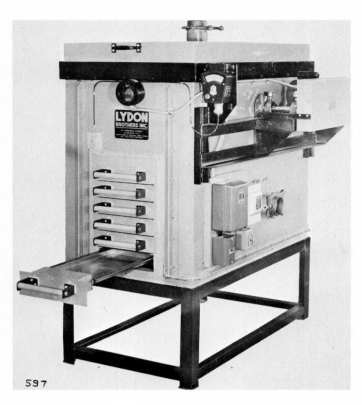

Figure 6–29   Various drying ovens were developed for conditioning the thermoplastics for injection molding. (Courtesy, Lydon Brothers, Hackensack, New Jersey.)

plastics raised the mold temperature into the steam zone and more. This brought on a variety of hot-cold heat exchangers to maintain a constant mold temperature. Highly sophisticated refrigeration units (fig. 6–28) followed for those products that required chilled molds for minimum cycles. Dow Therm and electrical heat have been used for material "chilling" of some of the high-temperature-melt materials. Local innovation solved all of these

problems nicely with vigor and fine engineering practice.

For conditioning thermoplastics, multiple drying pans were used initially, as shown in figure 6–29. This was followed by the continuous conditioning process (see fig. 6–27), which took material from the drums through driers in a continuous and enclosed system into the injection machiner hopper.

Mold-making practice followed the standards established for ther-

Figure 6–30   The development of this, the first standard mold base, by D-M-E Corporation in 1943 was a major contribution to all molding.

mosetting molds in each individual shop at the start. Runners ran radially from the central sprue. Edge gates or direct gates were used. Large gate areas were thought to be necessary, and elaborate degating fixtures were used in the early work. Pinpoint gating and the many special gating procedures that developed over the years introduced many unique mold design considerations. A major contribution to the molding of plastics was made by Mr. I. T. Quarnstrom, who founded D-M-E Corporation and marketed the first standard mold base (fig. 6–30) in 1943. Mr. Quarnstrom and his 1943 organization are shown in figure 6–31. The line of this organization has been expanded to include standard mold components for all possible mold parts.

Many ingenious fixtures were developed by the molders to mechanize injection molding and permit

Figure 6–31   Mr. I. T. Quarnstrom is shown at the left of the top row of this picture of the entire D-M-E organization in 1943 when the first standard mold base was made.

the operator to tend many machines. Such devices included weight-measuring units that locked the machine if the entire proper weight of the charge was not ejected after each cycle. Press-actuated wipers ensured removal of parts in some molds. Motor-driven and press-actuated unscrewing devices were developed at an early date to turn the cores when threads were molded in the pieces. Cam-actuated and hy-draulically actuated side cores facilitated the automatic operation of very complex pieces. A universal robot is shown in figure 6–32, which can be programmed to do a series of mechanical things.

Calendering

Calendering as used for coating webs was developed by Edwin M.

Figure 6–32 *Shown here is the first robot used in the molding of plastics. This machine is set to unload the pieces from the mold and unload them properly positioned in spray mask for painting. The machine then stacks the pieces on a pallet for drying and storage, with cardboard separators between each layer of pieces. It can be programmed to do any series of desired operations. (Courtesy, Unimation, Incorporated, Danbury, Connecticut.)*

Figure 6–33 *The Chaffee Calender as developed in 1836 for Roxbury India Rubber Company by Edwin M. Chaffee. Vinyls today are calendered in similar machines.*

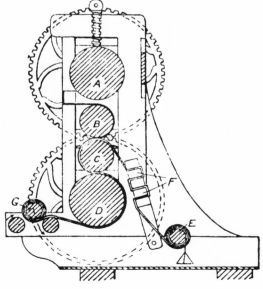

Chaffee in 1836 for the Roxbury India Rubber Company of Roxbury, Massachusetts. His device is shown in figure 6–33. It consisted of two rolls (18-in. diameter) in the top and bottom positions and two rolls (12-in. diameter) in the center position. Rubber was fed into the top nip where it was warmed and sheeted before being applied to the fabric in the middle nip. Present day calenders used for all thermoplastics follow this early design. Most of the early rubber goods have been converted to synthetic rubbers and to vinyl plastics.

Solvent Dip Molding

Solutions of plastics have been extensively used for shaped parts by the solvent molding or dipping process. In this method, a male mold shape is immersed in the cellulose acetate solution, where film is deposited on the surface. Vinyls and the acetates were molded in volume

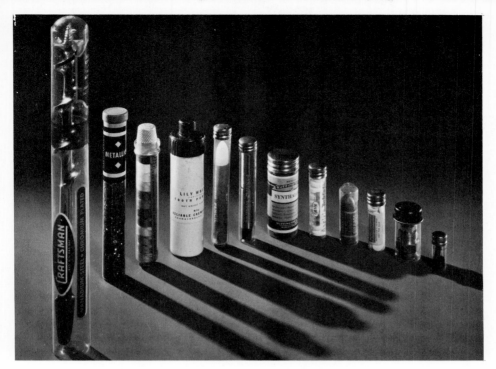

Figure 6–34   These solvent-dipped cellulose acetate containers were widely used before being displaced by injection-molded-and-blown containers. Celluplastic Corporation developed fully automatic machinery for these parts in the 1930s. (Courtesy, Hercules Powder Corporation, Wilmington, Delaware.)

by this process. Celluplastic Corporation produced a high volume of thin-walled containers (fig. 6–34) by the solvent dip molding process using proprietary automatic machinery.

Fluidized Bed and Electronic Spray Coatings

This technique was originated by Knapsack Griesheim AG of Frank-

furt, Germany, in 1953. Polymer Processes Incoporated acquired a license under the U.S. patents and introduced their Whirlclad Coating System to the United States in 1953 and have many licensees using the process. The fluidized bed-coating process is based on dipping a preheated metallic part into a bed of finely divided, dry plastics powder that melts and fuses on the metal parts. An alternate process developed in the early 1960s, called the Electrostatic Spray Process, works

Figure 6–35   Plastics collapsible tubes were introduced in the early 1950s.

on the principle that oppositely charged particles attract each other. The part to be coated is grounded and attracts the charged particle of the plastics. These processes are used for the thermosets and thermoplastics.

Plastics Tubes

Plastics tube containers with re-

movable caps as depicted in figure 6–35 were first developed in Switzerland in 1951 by Andre Strahm and his associates in Uni-Tubo S./A. (now Tuboplast S./A.) Precut lengths of tubing in the initial process are inserted on a mold mandrel and the neck shoulder end is welded to the tube while being formed in the injection mold. The incoming hot molten plastics bonds to the tubing. An alternate process was developed by Myron H. Downs in 1959. This process uses similar tubing but introduces a molten disc of plastics that welds to the tube and forms the neck-shoulder portion. F. E. Brown developed a third process in 1967 in which the neck-shoulder end is injection molded but does not bond to the tube. Frank Brown's process creates a mechanical grip to the tube and permits the use of low permeability tubing with polyolefins for the dispensing end. In all cases these tubes are capped, filled, and the end heat sealed.

Initially Plax had taken a license under Uni-Tubo patents but released it in 1952. Bradley Dewey picked up the Plax option and with Henry E. Griffith formed Bradley Container Company in Maynard, Massachusetts, in 1953. The Swiss machines had to be replaced with more rugged models that were built by Fred Prahl and associates; real production finally started in 1956 and the market grew very fast. Bradley Container was subsequently acquired by Olin-Mathieson, who sold it to American Can Company. Several others have followed into this large market to supplement the collapsible metal tubes with the transparent and colorful plastics.

The Downs process was further developed by Thatcher Glass Company to supplement their previous practice. The resulting production machine was licensed to Peerless Tube Company, Brockway Glass Company, and Tubed Products Company. Tubed Products Company also produces some blow-molded plastics tubes under the Maas 1965 patents. Many molders have tried to produce tubes directly

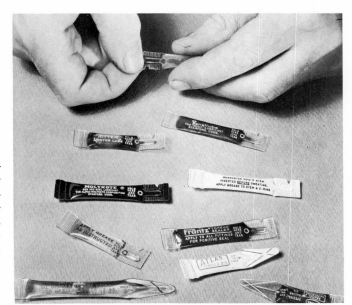

Figure 6–36   Sealed tubes of impermeable plastics material are widely accepted for the packaging of difficult pharmaceuticals and oils. (Courtesy, Unette Corporation, Livingston, New Jersey.)

by the injection molding process but found it difficult to form a thin concentric wall.

Plastics Tube Containers

Plastics tube containers (fig. 6–36) with sealed ends and tear-off opener were initiated by Unette Corporation in 1955. These units were filled and sealed in proprietary ma-

chinery for users of such special-purpose, single-use containers. Saran tubing is most often used because of its better protection against migration and loss of the contents.

Rotational Molding*

Rotational or rotomolding was practiced many years ago by metal

* Also known as slush molding.

molders under the name of slush casting. They poured a molten charge of metal in a mold, closed the orifice, and the rotating mold chilled a thin section of metal on the surface, making a hollow product. Chocolate rabbits, easter eggs, etc., have been molded by this process since the teens. This process was also used for the cast phenolics and Plastisols.

The rubber industry developed slush molding procedures for plasticized vinyls. Gloves, doll parts, packages, toys, etc., have been made for many years by this forerunner of rotational molding. In this original process, the Plastisol is poured into a hot mold, and, when a small portion gels on the mold surface, the excess is poured off; then the gel is cooled and stripped out of the mold. In contemporary rotational-molding procedures, the mold is enclosed in an oven where it rotates in

Figure 6–37 This early model of the McNeil Roto-Cast Machine shows the rotational moldings of play balls. (Courtesy, McNeil Akron Division of the McNeil Corporation.)

two planes while melting and distributing the plastic material on the mold surface. Cooling then finishes the cycle.

Contemporary rotational molding began to take shape in 1958 when finely divided low-density polyethylene was becoming available. Much of this development resulted from work by materials makers to open new resin markets. The McNeil Akron Division of McNeil Corporation (fig. 6–37) contrib-

uted greatly to the improvement and size enlargement of machinery.

The last decade has seen a great increase in the size and sophistication of rotational molded parts. Contemporary products include 4,000-gal tanks, phonograph cabinets, luggage, boats, ice chests, dashboards, heater ducts, etc. A double-walled shipping container 13 ft long, $2\frac{1}{2}$ ft square with $\frac{3}{16}$-in walls is used to transport 880 lb of military parts. Other contemporary

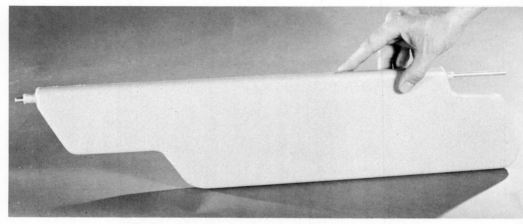

Figure 6–38 This air-cushioned safety sun visor for automobiles is rota-tionally molded from powdered polyethylene. It eliminates the need for hard board, padding, tempered wire frames, and metal tubing.

markets include boat fenders, store displays, chemical tanks, arm rests, sun visors (fig. 6–38), windshield blower ducts, battery jars, displays, snow-blower components, chair frames, and streetlight globes.

Thermoforming*

The keratin materials (chap. 1) horn, tortoiseshell, and hoof were thermoformed in the United States during the eighteenth century following methods used by the early Egyptians. The combmakers depended on thermoforming to shape their products. The forming of Celluloid sheets was handled by these combmakers in the same manner, and the larger Celluloid sheets enabled them to expand their product lines. The blow-molding process

* Also called vacuum forming and drape forming.

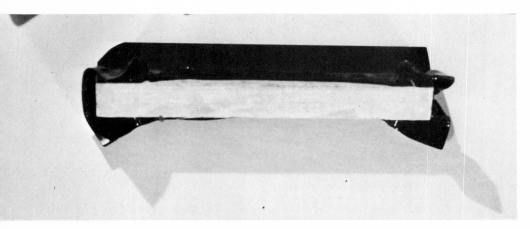

Figure 6–39   Cellulose nitrate was thermoformed over shaped wooden cores for the "patent" sharps keys of the decade following 1910.

(chap. 2) developed by Hyatt was in reality a thermoforming process since he used heat and pressure to shape a tube of cellulose nitrate in a die as illustrated in figure 8–25. Steam heating was used in this pioneering work because of the high flammability of Celluloid. Cellulose nitrate products were thermoformed into many products such as the piano sharp (fig. 6–39), which was shaped over a wood core.

Work done by Mr. J. J. Braund of the Coast and Geodetic Survey in the early 1930s to evaluate the feasibility of printing flat maps on thermoplastic sheets prior to forming them into three-dimensional relief maps started a new study from which developed much of our modern procedures. E. Bowman Stratton, Jr., as chief of the Materials Development Section for the Army Map Service, Corps of Engineers, developed this idea further in 1947 and produced highly successful

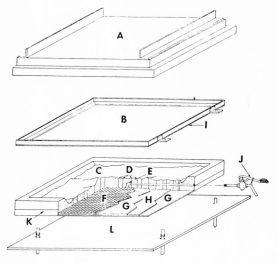

plastics three-dimensional maps. His process is shown in figure 6–40. He continued work on this process after leaving the Army Map Service. Thermoforming and skin packaging procedures with production equipment were developed by Stratton at Industrial Radiant Heat Corporation, Auto Vac Company and at Brown Machinery Company. The first public showing of skin-packaged products was made at the Hardware Manufacturers' Associa-

tion convention in Chicago, March 1954. A wide variety of commercial machines followed this pioneer work.

The Strauch machine (fig. 6–41) was one of the first if not the first of the fully automatic thermoforming machines. It was built by Clauss B. Strauch and purchased by Plax Corporation in 1938. All motions were mechanical and cam operated. It had all of the basic design features of the contemporary machines. A

Figure 6–41   This fully automatic thermoforming machine was developed by Clauss B. Strauch in the late 1930s. It produced in such high volume that it was impossible to keep it busy with the available markets and materials. A typical product was the cellulose acetate cigarette tip shown in figure 6–42.

chain belt with clips held the sheet as it indexed automatically from heating station to forming station and punchout station. The only material available at that time was cellulose acetate sheet. Typical production parts were christmas-tree spires, stars, cigarette tips (fig. 6–42), and ice-cube trays.

Many refinements were added to this machine and to the art of thermoforming by Ludwig Fehrenbach.

L. H. Pfohl of Design Center in New York started work on a blister package in 1938. He softened acetate sheet on a gas-heated plate and, while it was held in a frame, shaped it in a press.

The first blister package by this process is credited to Becton Dickinson in 1942.

Gustave Borkland of Borkland Laboratories in Marion, Indiana, initiated a process for making preprinted plastics Christmas cards with three-dimensional effects.

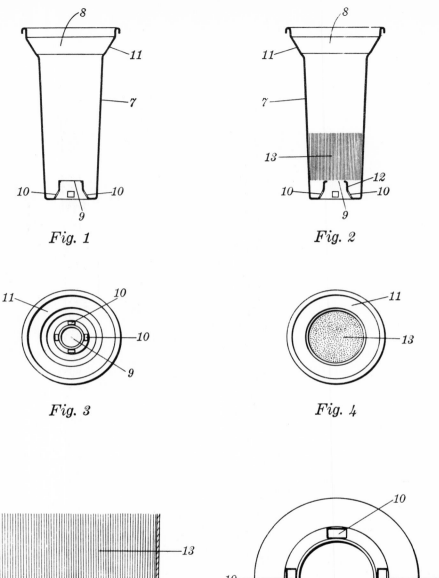

Fig. 1

Fig. 2

Fig. 3

Fig. 4

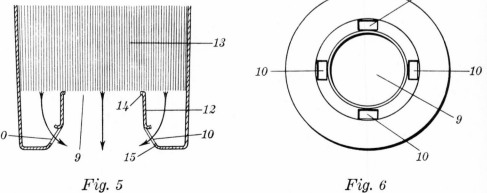

Fig. 5

Fig. 6

Figure 6-42 One of the early thermoformed products in the era after Hyatt was this cigarette tip for which Clauss B. Strauch took out a patent.

*Figure 6–43 Bomber noses of Plexiglas acrylic resin were thermoformed
for war planes in the early 1940s. (Courtesy, Röhm and Haas Company,
Philadelphia, Pennsylvania.)*

A milestone in the tremendous growth of the thermoformed plastics for packaging was the introduction by Kraft foods of individual jelly servings in vinyl packages that were vacuumed formed, mechanically filled, and sealed.

The thermoforming of aircraft turrets, astrodomes, and window and cockpit enclosures was developed to a fine art in World War II. The acrylic sheets were heated in ovens up to 250°F, at which point they were soft and pliable and could be shaped over a cloth-covered form. Other sheets were stretched and clamped over a vacuum pot, where the sheet was drawn by vacuum into the desired shape as shown in figure 6–43.

Dr. Donald S. Frederick of Röhm and Haas Company was given the 1942 Hyatt Award for his work in the adaptation of large sections of methyl methacrylate to the manufacture of military aircraft.

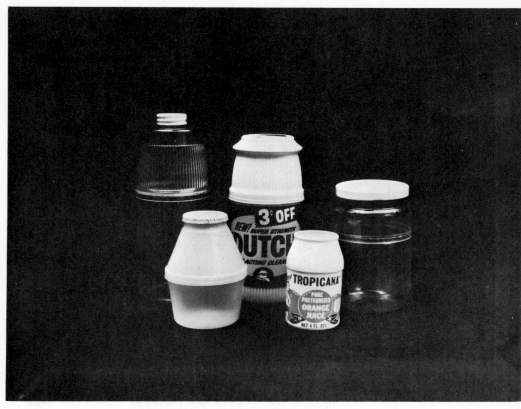

Figure 6–44   These thermoformed containers are made in two pieces and then are spin welded. Most thermoplastics materials can be used for this process. (Courtesy, Brown Machine Division of Koehring Company, Beaverton, Michigan.)

Fernplas Company started their work on flexible plastics containers in the early 1930s, and this led them to the two-piece construction as illustrated in figure 8–26. The materials available to Ferngren at that time handicapped the realization of economical production procedures for the two-piece thermoformed bottle. The bottles shown in figure 6–44 were produced in 1970 by the latest thermoforming and spin-welding techniques (fig. 6–45) —

40 years after Ferngren's original conception of the thermoformed bottle.

Many types of thermoforming machines were developed during the last two decades as the variety of materials and properties developed. Blister packages, bottles, display packaging, pallet-protective covering, meat trays, product components, and structural parts all gained special machine developments to expand their applications.

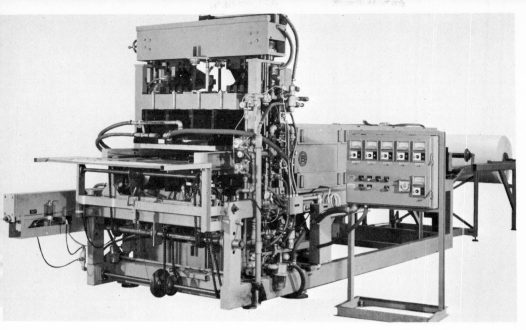

Figure 6–45 The *1970* thermoforming machine is a very sophisticated device when compared with its Strauch predecessor. (Courtesy, Brown Machine Division of Koehring Company, Beaverton, Michigan.)

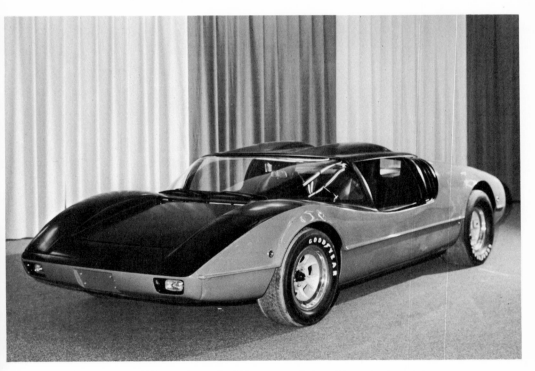

Figure 6–46 Thermoformed ABS was used for this experimental car body in *1970*, designed by Borg-Warner's Centaur Engineering facility. This is now considered by some to be the plastics car body of the future. (Courtesy, Borg-Warner Corporation.)

Figure 6–47   An early application for Styrofoam expanded styrene was for thermal insulation in cold-storage rooms.

The experimental automobile body shown in figure 6–46 illustrates the 1970 trend to larger segments of thermoformed plastics where they will replace metal in many high-volume applications.

The Expanded Plastics

The foamed, expanded, or bubble plastics were first used during World War II for filling the buoys that suspended the submarine nets so they would not sink when riddled with machine-gun bullets. Phenolic foam from General Electric Company and Styrofoam from Dow Chemical Company were used for this process. The invention of phenolic foam was easy; preventing its formation had always been a problem in making phenolic resin. Phenolic foams are produced by the concurrent condensation and foaming of the resin. Styrene foam

was invented in Sweden by Magnus, and in the United States it was first produced by Dow Chemical Company (fig. 6–47).

Two basic types of foam have been developed. Light foam weighs 6 to 15 lb per cu ft or three-quarters air and one-quarter resin. Structural foam is one-quarter air and three-quarters resin and weighs 45 lb per cu ft. The structural foams are largely made by the Union Carbide process.

The armed services were quick to find ways to use these new lightweight structural materials for the replacement of balsa wood in aircraft. Stiff and lightweight aircraft skins with polyester-glass surfaces and foamed plastics interiors were produced in the early stages of World War II. These structures like I-beams found many wartime applications. Phenolic, styrene, cellulose acetate, polyvinyl chloride, polyvinyl formal, and synthetic

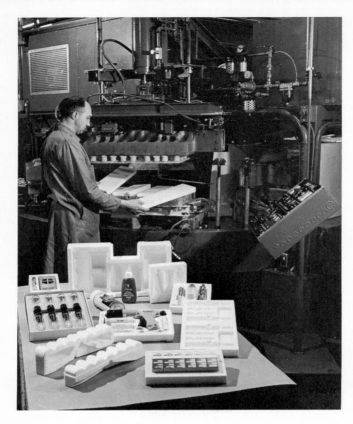

rubber foams were used in World War II.

As thermal insulating materials, the expanded plastics moved into the structural, refrigeration, and other such markets where they could be applied by pouring into the void. In the postwar years, various molding methods and compounds were developed by several material makers to facilitate high-volume molding, extrusion, and casting of the expanded plastics.

Styrene foam products are commonly produced from expandable beads of styrene produced in several ways. Exposure of these beads to heat results in expansion (fig. 6–48) from increased vapor pressure of the blowing agent, forming discontinuous cells. Since this art is still relatively new and the patent literature is incomplete, it is not possible to give a real historical study of the foam development.

A commonly used original proc-

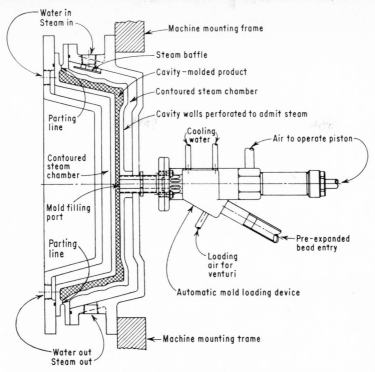

Figure labels (top to bottom, left to right):
Water in
Steam in
Machine mounting frame
Steam baffle
Cavity-molded product
Contoured steam chamber
Cavity walls perforated to admit steam
Parting line
Cooling water
Air to operate piston
Contoured steam chamber
Mold filling port
Parting line
Pre-expanded bead entry
Loading air for venturi
Automatic mold loading device
Machine mounting frame
Water out
Steam out

Figure 6–49 Cross-section of a typical 1960 mold for expandable bead automatic foam molding. Also shown, attached to the right of the mold, is the device for automatically loading the preexpanded beads into the cavity. (Courtesy, **Modern Plastics.***)*

ess for molding expanded plastics products is explained in figure 6–49 as developed by Frank H. Lambert in 1963. This process was widely used for disposable packages, ice buckets, etc.

Haskon Division of Hercules Incorporated developed a process for compression molding expanded polypropylene. They preblended the polypropylene with blowing and crosslinking agents in a screw extruder prior to the compression molding process. A pressure vessel is used for the preblending followed by quick cooling after molding in the process developed by Du Pont.

Phillips Petroleum Company has developed a method based on the Engel extruder (fig. 8–16) for plasticizing the input. Blowing agents and pigments are injected into the extruder barrel. Conventional resin pellets and scrap may be used in this method, which has numerous advantages for furniture molding.

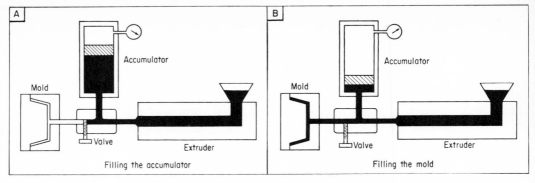

Figure 6–50 *The Union Carbide process invented in 1963 by Richard G. Angell, Jr., uses an extruder to feed the plasticized resin and blowing agent into the accumulator. When the predetermined charge is reached, an accumulator plunger fills the mold, where expansion takes place.*

Figure 6–51 *This 1970 structural foam molding press, invented by Richard Angell, Jr., was developed by Williams, White and Company for use in conjunction with the Union Carbide process. (Courtesy, Williams, White and Company.)*

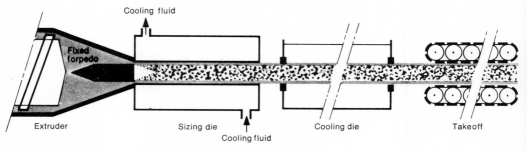

Figure 6–52 *The Celka or French process uses a fixed torpedo at the extruder exit to create a hollow space into which the extrudate expands. Skin is formed by the cooled sizing die.*

Figure 6–53   The urethane foams may be applied on the spot with simple applicators. (Courtesy, Insta-Foam Products Incorporated, Addison, Illinois.)

The Union Carbide process (fig. 6–50) was developed originally for polypropylene and high-density polyethylene. Resin and blowing agents are plasticized in an extruder with the extrudate going into an accumulator where it is held under pressure, above its foaming temperature, and then discharged into the mold by the opening of a valve. The high-pressure mix expands rapidly in the no-pressure mold. A contemporary machine is shown in figure 6–51.

The Celka or French process for structural foam has a torpedo at the end of the screw, which creates a void at its other end where the compound expands in the cooled sizing die, as shown in figure 6–52.

Allied Chemical Company introduced another process that is adaptable to screw injection machines. It permits the excess foamed material to return from the mold for injection in the following shot. Several compounds have been developed that include blowing agents so that the foamed product may be molded in conventional injection-molding machines.

The urethane foams (fig. 6–53) may be formed by spraying or pouring the two components into the mold; this is particularly useful for insulating structures. For injection molding or extrusion, valves will meter and blend the components as they flow into the mold.

7

The Thermoplastics Explosion

As a result of the plastics industry's having "taken over" the chemical and petrochemical industries, the following initial historical sketch* is given.

The early colonists brought the chemistry as practiced in England. They exported potashes to England in 1608, one year after the founding of Jamestown. Massachusetts gave

* Data on early chemical industry and the rubber program were extracted from "The Rise of the United States Chemical Industry," *Chemical Week*, November 1968.

Samuel Winslow a monopoly to collect and sell salt in 1641, and the first United States patent was issued to Samuel Hopkins on an improved potash kettle. The initial chemical operations in the United States produced products from natural materials—potash from wood ashes, glass from sand, potash, and salt. To get salt, John Sears built a vat 100 ft by 10 ft, filled by a windmill with sea water. He learned selective crystallization in order to get sodium chloride, epsom, and Glauber's salts. Gunpowder was made from 10% sulphur, 15% charcoal, and 75% saltpeter. In 1642 the General Court of Massachusetts ordered that every plantation should erect a 20 ft by 20 ft house "to make saltpeter from the urine of men, beasts, goates, henns, hogs and horsedung." These materials were mixed with weeds, slaughterhouse offal, limestone, and ashes. After due time and leaching, saltpeter was crystallized therefrom.

In 1800, E. I. du Pont de Nemours built a black powder mill on Brandywine Creek in Delaware. Another refugee from the French Revolution, Nicholas Lennig, built a large sulfuric acid plant in Philadelphia in 1830. Eventually this became part of Röhm and Haas Company. Charles Lennig, a son of Nicholas Lennig, was a party to the

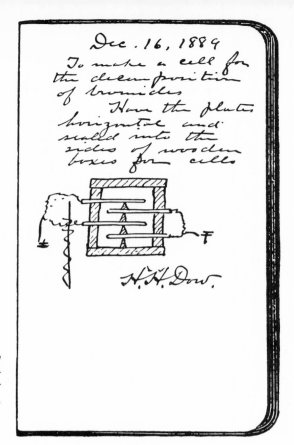

Dec. 16, 1889

Figure 7–1   Herbert H. Dow invented this novel bromine-extraction process shortly after leaving college.

formation of Pennsylvania Salt Manufacturing Company in 1850. In 1872 the Manufacturing Chemists' Association was formed, followed by the American Chemical Society in 1876. In 1889, Herbert Dow, fresh out of college, invented a bromine extraction process (fig. 7–1), which started the Dow Chemical Company. To produce calcium cyanamide fertilizer, Frank Washburn, a civil engineer, started Amer-ican Cyanamid Company in 1907. The Lithoid Corporation, founded in 1886 to make pyroxylin plastics in Newburyport, Massachusetts, was reorganized in 1870 as Fiberloid Corporation and in 1930 became the Plastics Division of Monsanto Chemical Company.

At the beginning of World War I, Fries' Brothers, Heyden, Monsanto, and Benzol Products were making salicylic and benzoic acids, chlorsulfuric acid, aniline oil, and nitrobenzene. Schoellkopf was the only maker of intermediates at that time. The plants built to meet the requirements of World War I eliminated our dependence on foreign sources of supply and built up tremendously the United States chemical industry.

Du Pont established its first research laboratory in 1903. As a result of the antitrust litigation which broke up Du Pont, some smaller companies were formed. Atlas and Hercules were split off from Du Pont by the 1911 antitrust ruling and moved into nonexplosive markets. The Mellon Research Institute founded in 1913 brought together Union Carbide and Prest-O-Lite into Union Carbide and Carbon Chemicals Corporation along with Electromet, National Carbon, and Linde Air Products. This company absorbed Bakelite Corporation.

In 1916, Carleton Ellis initiated the petrochemical industry through his discovery of a method for making isopropyl alcohol from cracked refinery gases by the absorption of the olefins in sulfuric acid. The isopropyl alcohol was then converted into acetone, and this process was taken up by Standard Oil Company of New Jersey and by Carbide and Carbon Chemicals Company.

Commercial Solvents, organized in 1919, produced butyl alcohol and acetone by the Weizmann fermentation process and later changed to synthetic butonol for automotive lacquer. They subsequently turned to synthetic methanol. This was the big move to the synthetics that started the downfall of wood distillation at Du Pont, Shawinigan, Niacet, Cities Service, with others joining the parade.

In chapter 2 is told the story of how the use of cellulosics initiated the synthetic plastics and fabrics. At this point plastics became serious business with an obvious future. The explosive growth of the phenolics and the cellulosics in the 1920s and 1930s started the snowball of plastics that has taken over these basic industries.

Any record of thermoplastics history is an exciting recital of business acumen, shrewd marketing, and the development of molds, machinery, and polymers. Hyatt's semisynthetic Celluloid replaced such natural products as ivory, shellac, shell, horn, hoof, amber, and gutta-percha and created its own boundless new markets. Celluloid continued to be an important material until 1950, although its growth stopped soon after 1930 when cellulose acetate started its move into the mass markets and triggered the thermoplastics explosion.

Injection molding and the acetate plastics were responsible for the next giant stride of the plastics industry. Industry's quick adoption and use of the acetates expedited in-

Figure 7–2   This 1857 daguerreotype case was used as a jewelry case and may well be the first plastics package.

jection machine developments and this was further inspired and followed by many new basic thermoplastics. The thermoplastics of today have taken over more than two-thirds of the total plastics market in 25 years.

Thermoplastic Materials

Shellac, our first moldable natural plastics material, is a thermoplastic consisting of various hydroxy acids partially combined with each other as lactones and anhydrides. Shellac compounds were initially compression molded (fig. 7–2) and then injection molded as that process developed. Shellac continued to be an important basic material for molded and laminated products through World War II. Material shortages during the war introduced the salvage of old shellac phonograph records as a source of molding material.

Early studies in Europe had dis-

Figure 7–3   This cellulose acetate plant at Cumberland, Maryland, was under construction as shown here at the end of World War I. It was being built by the Dreyfus brothers for the War Production Board for airplane fabric dopes. In the 1920s the plant was completed for the making of rayon. (Courtesy, Celanese Corporation of America, Cumberland, Maryland.)

closed cellulose acetate (chap. 2) in 1844. A. Eichengrün and T. Becker revived interest with their 1903 German patents. G. W. Miles, an American chemist, in 1905 developed cellulose acetate fibers by a new process that proved to be uneconomical at the time. George Eastman, who had introduced transparent roll film on a nitrocellulose support in 1889, introduced cellulose acetate safety film in 1908. Oriented styrene and polyester ultimately became photo film also.

The Swiss brothers, Henri and Camille Dreyfus, initiated commercial production of acetate lacquers and film in Basle in 1910. They also had, in cellulose acetate, the right answer for stiffening airplane fabric and built a plant for the British which was later to become British Celanese Limited. The extended needs of World War I caused our War Industries Board to ask Camille Dreyfus to start an American plant (fig. 7–3), the American Cellulose and Chemical Company,

Figure 7–4   Rayon fiber making is shown here in the early 1920s at the Dreyfus plant, American Cellulose and Chemical Manufacturing Company, Limited, in Cumberland, Maryland. (Courtesy, Celanese Corporation of America, Cumberland, Maryland.)

Limited, which was incomplete at the time of the 1918 Armistice and became Celanese Corporation of America in 1927. That initial cellulose acetate plant was finished and started making acetate yarn and fibers (fig. 7–4) in 1925. Other American rayon operations at that time included Cellulose Products Company, American Viscose Corporation, DuPont Fibersilk Company, and Lustron, Incorporated, with pioneer workers Arthur D. Little, William Walker, Gustavus

Esselen, Charles A. Ernst, and Harry Mark contributing. The rights to the Arthur D. Little patents on films were acquired by Eastman Kodak Company, and the fibers went to Celanese Corporation of America. Hercules entered this field in 1938 to expand the uses for its cellulosics.

Celluloid Corporation started production of its Lumarith CA sheets, rods, and tubes in 1927. The May 1929 issue of *Molded Products* features this story: "New White

Light-Colored Powder. The Celluloid Corporation announces the arrival of a workable cellulose acetate molding powder which the industry has awaited for years." Compression molding with 40-lb steam pressure and 2-tons molding pressure was suggested. This arrival of cellulose acetate was a most important event. Noteworthy is the fact that urea resin was also announced in 1929, giving additional color to the thermosets.

These Lumarith materials were of excellent quality, but production procedures followed the 200-lb-batch Celluloid solvent process (chap. 8) methods which were time consuming and produced a costly product. The early Lumarith market did not develop fast as a result of these handicaps. Celanese Corporation of America, presided over by Camille Dreyfus (fig. 7–5), merged Celluloid into Celanese, and the present plastics operations are

Figure 7–5   Dr. Camille Dreyfus with his brother Henri did development work at an early date on cellulose acetate. Dr. Dreyfus put together Celluloid Corporation and Celanese Corporation of America, which contributed tremendously to the growth of the plastics industry.

known as Celanese Plastics Company. It is significant that the acetate phase of the plastics industry was fathered by the tremendously successful film and synthetic fabric industry. Celanese Plastics Company ended its cellulose acetate production in 1970.

The Grotelite Company* imported 12 Bucholz injection machines in 1922 and thus became the

* See chapter 6 for more details on early injection molding.

first American injection molder. They were licensed to build this machine in 1923.

In 1929, Eastman Kodak Company assigned their enlarged film production of cellulose acetate to Tennessee Eastman Company, which went into production in 1930. Their market research program investigated textile yarns and molded plastics. The Tennessee Eastman method produced acetate yarn in 1931 and moldable plastics granules in 1932. Their memorable marketing drive to enter the molded plastics field was started by Spencer E. Palmer and John Slater, who initiated the big swing to thermoplastics. Theirs was the job of educating the old compression molders to accept the long heat and chill cycle of compression molding for thermoplastics and, later, of selling the injection process. As was experienced by Dr. Baekeland in his introductory work, they made better progress with newcomers than they did with the older molders. Early sales included the Bronson fishing reel ends and Owens toothbrush handles. A concentrated drive on the automobile industry resulted in variegated knobs for the 1934 Oldsmobile, and the rest of the industry followed quickly. The acetate steering wheel on Chevrolet as an attractive accessory was followed in 1936 as standard equipment on the Hudson, and its success

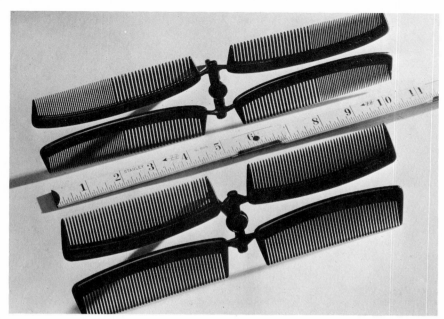

Figure 7–6   Combs, which played such an important part in the plastics industry, were among the first products to be injection molded.

touched off the big automotive acceptance. Combs (fig. 7–6) were a high-volume early product of the injection molders.

Celluloid reported injection studies in 1927 and Du Pont in 1931. Celluloid applied for a patent on a screw preplasticizer system in 1933 but did little with the idea. The Eckert-Ziegler and Isoma Automat injection machines had been imported since 1932 by several molders. Isoma was first to use the

Gastrow torpedo. The HPM hand injection-molding machine of 1931 stimulated more interest in injection molding and fostered many do-it-yourself projects. Tennessee Eastman commissioned Pack Morin Incorporated to design an injection machine, which was made for them by E. W. Bliss Company in 1934, becoming the first large-capacity injection machine. HPM brought out its first standard injection machine in 1935, and the 1936 Reed-

Prentice 10-D-8 was the most popular machine of all time (see chap. 6). Index Machinery Corporation offered a clock-controlled model.

The year 1936 is considered to be the turning point when the adolescent thermoplastics were recognized and started to mature as basic materials for industry. The 1936 Encyclopedia listed the following plastics materials: phenolic, urea, pyroxylin, phenolic cast resins, vinyl chloride, vinyl acetate, vinyl chloride-acetate, casein, ethyl cellulose, cellulose acetate, acrylic, furfural formaldehyde, cumarone, polystyrene, synthetic rubber, and shellac. It is interesting to note that only 5 makers of injection machines were publicized in 1936, and the machines were called extrusion molding presses.

The May 1936 issue of *Modern Plastics* carried a feature story, "What Is This Injection Molding?" by Vincent D. Hery wherein he

Figure 7–7   Extruded cellulose acetate table trim by Wernco was an initial civilian application for extruded thermoplastics. The tableware handles were injection molded. (Courtesy, Wernco.)

states, "Because of recent developments, it is now the opinion of many leaders of the plastics molding field that injection molding of thermoplastics will soon find its rightful place in the molding industry." This was in 1936! It found its place all right and almost took over! Other machine and material makers were quick to follow, and the race was on. Ethyl cellulose in 1936, cellulose propionate and methyl methacrylate in 1937, and butyrate

in 1938 opened still more markets for the injection molders. Hercules started production of cellulose acetate molding compound and ethyl cellulose in 1939. Detroit Macoid pioneered acetate extrusion (fig. 7–7) for furniture and automobile trim and, from this, profile and rod extrusion grew at a fast pace. Tennessee Eastman defended the industry and won, with Hyatt's prior art, against the Eichengrün and Bucholz machine patents on injection

Figure 7–8   This cellulose acetate shoe heel was first in the field. It was nailable and had a removable tap.

Figure 7–9   Army bugles were injection molded of cellulose acetate by Elmer E. Mills Company in the early days of World War II.

Figure 7–10   This army canteen was injection molded of ethyl cellulose early in World War II. Blow molded products are used today.

molding. The acetates and injection molding were now firmly established and continued to grow until they were beat out in many of the early markets by polystyrene and subsequent materials.

The ladies' shoe heel started with Du Pont's Pyraheel, a wood heel covered with cellulose nitrate. In 1938, Pearson Heel Company, working with General Electric Company, introduced a nailable cellulose acetate heel (fig. 7–8) with a snap-in tap. This became a high-style item for a few years. The almost universal adoption of plastics for ladies' shoe heels came along in the early 1950s with high-impact styrene, which made this a more economical application.

World War II found many interesting uses for the cellulosics, such as ammunition feed rollers, visors, scabbards, bullet-core tips, tax tokens, urinals, bugles (fig. 7–9), canteens (fig. 7–10), instrument

Figure 7–11 Double-shot molding was a boon to makers of business machines and instruments.

and machine components (fig. 7–11), and gas masks (fig. 7–12).

Self-extinguishing grades of cellulose acetate enabled it to be used in some applications (fig. 7–13) previously limited to the thermosets.

The story of Cellophane goes back to the early 1900s when Dr. Jacques Edwin Brandenberger, a Swiss textile chemist, conceived the idea of making tablecloths impervious to dirt and stains by treating them with a cellulose solution. He produced a smooth, sparkling cloth, but one that was stiff and brittle. Later, he made a thin sheet of transparent film and applied it to the cloth.

The tablecloth project was not a commercial success, but the thin transparent film offered bright possibilities. By 1911, Dr. Brandenberger had designed a machine for production of the film, which he named Cellophane. Shortly, the first

Figure 7–12   The plastics eyepieces, Y-tube, and speaking apparatus in this officer's gas mask were injection molded of cellulose acetate in World War II.

Figure 7–13   This A. C. Gilbert Whirlbeater housing was the first hand-held device to gain Underwriter approval. A self-extinguishing cellulose acetate was used. Aircraft trim tab motors were converted to this beater after World War II in a swords-to-plowshare action.

Figure 7–14 Development of the first successful moistureproof cellophane film is reenacted here by Du Pont's late Dr. Hale Charch. Charch and his associates developed the new film after they had mixed 2,500 formulas and made hundreds of tests. The large bag on Charch's left held water for weeks, while other control bags showed evaporation in a few days. (Courtesy, E. I. du Pont de Nemours and Company, Wilmington, Delaware.)

Cellophane plant was built in France by La Cellophane Société Anonyme. It was from La Cellophane that Du Pont acquired the American rights to the Brandenberger process in 1923. The first sheet of Du Pont Cellophane came off the casting machine in Buffalo, New York, in the spring of 1924. The most important milestone in history of Cellophane was the development in 1927 of a way to make a new product, moistureproof Cellophane. This new film, developed by Du Pont's Dr. Hale Charch and associates (fig. 7–14), opened up a whole new field of uses for Cellophane, particularly in wrapping foods and tobaccos. American Viscose Corporation and Olin Industries were licensed later to produce Cellophane to avoid monopoly problems.

The vinyl resins were discovered by Regnault in 1838. Bauman produced solids in 1872, and Ostromi-

Figure 7–15   Vinyl records were first introduced at A Century of Progress, the world's fair, in 1933.

sensky patented rubberlike vinyl products in 1912. An initial story was published in the February 1930 *Plastics* entitled, "Vinyl Ester Resins Now Commercially Available." Carbide and Carbon Chemicals Corporation first showed publicly their Vinylite products (fig. 7–15) at Chicago's 1933 World's Fair, A Century of Progress. Floor tiles at the Fair proved their color stability and wear resistance.

Intensive research by Carbide and Carbon Chemicals Corporation resulted in the first successful commercial development of the vinyl materials in 1927. It had been recognized from the results of extensive market research that something had resulted unlike any other material in the field of plastics. Dr. Waldo Semon of Goodrich is credited with important developments in the plasticization of PVC for molding compounds in the period from 1927 to 1933. Dr. Semon

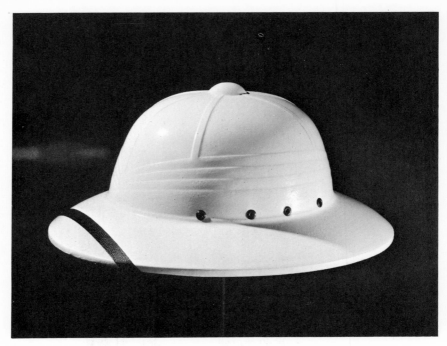

Figure 7–16   This 1935 Hawley Tropper had a ther-moformed vinyl film on the surface. Hawley Products pioneered the pulp preform-molding process to make loudspeaker diaphragms.

worked on the plastisols also and contributed to their production and use. The earliest applications from 1927 to 1931 included their use as a base for lacquers and impregnating solutions, moistureproofing compounds for wallboard, gypsum, and plaster. Toothbrush handles, picture frames, clock cases, cigarette boxes and dentures were formed from molding materials made of vinyl resins. They provided previously impossible colors. By 1931,

vinyl compounds were made for molding radio transcription records. As a result of their extremely low noise level and high mechanical strength, such records continue to be used today. Records for broadcasting studios and the highest quality transcriptions for home use are still made from vinyl compounds. Coatings for metal foil to be used in food packaging followed soon after. The Hawley Tropper (fig. 7–16) was coated with a thermoformed vinyl sheet in 1934. In 1934 paper coated with vinyl plastics was used for bottle-cap liners, providing an impervious seal, still unequaled for food, pharmaceutical, and cosmetic containers.

Rigid sheets made from vinyl plastics were first used commercially during 1934 as sun visors for automobiles. In 1936 they were used for the familiar plastics binding of books. Work started for Carbide and Carbon in 1934 by Arthur

K. Doolittle and G. M. Powell on the development of vinyl chloride copolymer resins for solution coatings resulted in the 1936 vinyl lining for beer cans and many other high-volume applications. Viscosity composition relationship charts for a variety of combinations of solvents and nonsolvents in the organosol field were published as a result of work done by Carbide research workers in South Charleston. G. M. Powell reported his subsequent work on resin plasticizer pastes, which John Sutter named Plastisols. A very important initial application for these new Plastisols was for filling the voids between conductors in the degaussing cables put around ships to avoid detonating magnetic mines. G. M. Powell was given the 1950 Hyatt Award for his work in introducing the vinyl dispersion resins.

Automobile safety glass took a big step forward in 1936 when a new type of vinyl plastic—polyvinyl butyral—was used for bonding two layers of glass. Unlike previous adhesives, it retained its transparency and toughness in all types of conditions. The safety glass invention of Union Carbide's H. F. Robertson provided a specific plasticizer for a resin having a specific butyral and acetate content. During the next year, vinyl plastics were adapted to the then revolutionary injection molding process.

Figure 7–17   Extruded vinyl gaskets remained permanently flexible and replaced many rubber parts.

During these years the development of the elastomeric types of vinyl plastics was continuing. By the addition of high-boiling solvents to the rigid materials, a rubberlike compound was obtained that retained most of the important properties of the rigid materials and yet took on new properties that were to be of extreme importance. Wire and cable with vinyl insulation proved to be superior to rubber in many ways. Vinyl chloride is self-extinguishing. Flamenol vinyl chloride insulation produced by General Electric Company solved many old wire-insulation problems, giving flexibility, oil resistance, self-extinguishing properties, age retention of elasticity, as well as dielectric strength. Molded and extruded parts were made that were permanently flexible (fig. 7–17) and yet had marked resistance to water, oils, and chemicals. John Reilly and Ralph Wiley of Dow Chemical

Company introduced vinylidene chloride (Saran) in 1940. Ted Sloan revolutionized the packaging of meat at Oscar Meyer and Company with Saran. Made into film and sheeting, the vinyl plastics were used for shower curtains, raincoats, belts, suspenders, and shoes. S. Buchsbaum and Company introduced in 1939 the vinyl elastic plastics with their Elasti-Glass suspenders, belts, and garters. Vinyl resins were also contributors to the new flash bulb developed by General Electric. The vinyl adhesives bonded the fine flash wires to the glass and also added shatter resistance to the bulb. Vinyl resins were then being used advantageously for chemically resistant coatings in collapsible tubes. Vinyl raincoats (fig. 7–18) were produced for the army in 1942. Plastisol dolls (fig. 7–19) were introduced by Horsman Dolls Incorporated in 1946; the plastigels (fig. 7–20) came along in 1952.

Figure 7–18 *This army raincoat was water-proofed with vinyl plastics. (Courtesy, Chemical Division, The B. F. Goodrich Company.)*

Figure 7–19 *Following World War II, the plastisols were used to make very realistic dolls.*

Figure 7–20 *The vinyl plastigels can be worked by hand and hardened by a 350°F oven cure. They retain very minute details without distortion.*

Figure 7-21 This grained and textured "rosewood" is produced by laminating vinyl sheeting to a molded surface. (Courtesy, Masland Duraleather Company.)

Plastigels are shaped at room temperature with hard pressure and then are oven hardened at 350° F. Vinyl-coated milk cartons were introduced by Sealrite Company in 1950, and vinyl liners for swimming pools were initiated by United States Fiber and Plastics and by Bilnor Corporation in 1952. Vinyl laminates (fig. 7-21) became very popular materials in the 1960s. The vinyl bottle (see chap. 8), developed in the midsixties opened mar-

kets that could not be served by the polyolefins (fig. 7-22).

A Kansas City physician-chemist, Dr. J. C. Patrick, discovered in 1923 that the thermoplastic chemical reaction product of ethylene dichloride and sodium polysulfide could be reacted to form a synthetic rubber. The Thiokol Corporation was formed in 1929 to make this product. Neoprene, which is a polymer of chloroprene, was developed in the Du Pont laboratories with com-

Figure 7–22 This blow-molded bottle was the first commercial application of a PVC container for oxygen-sensitive applications.

mercial production starting in 1931. It was first sold under the name of DuPrene.

Dr. Otto Röhm did his original work in 1901 on acrylic resins in Germany. Röhm and Haas Company first manufactured Plexiglas acrylic resins here in 1931 for coatings and safety glass. Sheet materials followed in 1936 and molding compounds in 1937. Du Pont entered this field in 1937 offering Lucite acrylic sheets and molding powders. Daniel E. Strain developed acrylic technology at Du Pont. The war placed heavy demands on the acrylic producers for aircraft windows and enclosures. Acrylic dentures were immediately accepted because of their more realistic color, natural texture and the ease of casting them in plaster molds. The H. D. Justi Company introduced their line of acrylic replacement teeth in 1946. The automotive industry was soon to use the acrylics because of

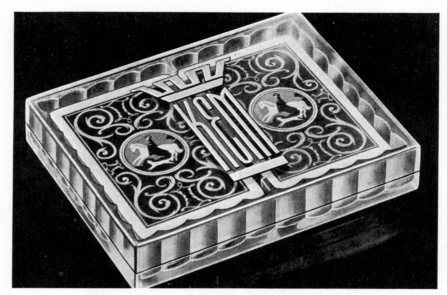

Figure 7–23   This 1937 Kem plastics playing-card box was compression molded of Lucite in beryllium copper molds. This package pioneered three-dimensional lettering in clear plastics and sparked the production of many automobile horn buttons and decorative insignia.

their excellent weatherability and their coloring potential. These features also gave the acrylics a fast growing market for packaging, display, and optical products. The Kem playing-card box (fig. 7–23) of 1937 set a new high in plastics packaging to be followed by Elgin National Watch Company display packages and others. Acrylic paints are easy to use and that, plus their long life, gave them a big slice of the paint market.

Styrene was well known in 1850, and it was synthesized by M. Berthelot in 1866. Staudinger had revived interest during the twenties and suggested that a more appropriate name might be metastyrene. In preparation for rubber shortages in wartime, I. G. Farben concentrated on Buna-S rubber and initial production of styrene in volume was started in 1929 in Germany. It was injection molded by Dynamit AG. Naugatuck Chemical Company did

Figure 7–24 R. R. Dreisbach of Dow Chemical Company, along with Dr. John J. Grebe, was given the 1947 Hyatt Award for the production of pure styrene and its polymerization.

initial work in 1925 but was limited by many handicaps. Styrene was marketed in 1936 under the trade name of Victron by Naugatuck Chemical Company. Some styrene was being imported from Germany at that time under the trade names of Trolitul and Resoglaz. Marvelyn Dental Corporation marketed styrene as a denture material called Victron.

The Dow study of styrene started in 1930, and in 1934 R. R. Dreis-

bach (fig. 7–24) developed a feasible process. Dr. Dreisbach and Dr. Grebe were given the 1947 Hyatt Award for their contribution to the production of pure styrene and its polymerization. Great credit is due Dr. Willard H. Dow who backed up the faith of his people who were sure they had a winner and were willing to make it go—at the bottom of the great depression. Marketing started on a small scale in 1937. In 1938, the plastics industry

Figure 7–25   One of the most spectacular achievements of the plastics industry was its contribution to the GR-S rubbermaking program during World War II. Shown here is a batch of synthetic rubber rolling from the intimate maze of tanks at Goodyear's wartime plant.

showed a net loss of 0.5% on total molded product sales of $46,000,-000. More than 100 of the new injection machines were operating with only a few having a top capacity of 9 ounces. In 1938 styrene production was 190,000 lb, and this rose to 1¼ billion lb per year in 25 years. Styrene now has full commodity status along with the polyolefins and vinyl plastics. During World War II, styrene monomer was essential to wartime rubber, and the Dow process was taken over by the government with plants being operated by Dow, Union Carbide, Koppers, and Monsanto.

One of the most spectacular achievements of the plastics industry was the wartime GR-S rubber program. Working with the rubber fabricators, gigantic rubber production (fig. 7–25) was achieved in three short years to supply the needs of the United States and its allies after the Japanese took over the Far East.

Even before the Japanese attack,

a development program was approved, calling for four plants producing 10,000 long tons per year to make styrene-butadiene rubber. Even this seemed a bold step at the time, since, although the rubber had been made, output in 1942 was less than 4,000 long tons.

After Pearl Harbor, it was decided to triple the capacity of each plant, making a total of 120,000 tons per year. As the Japanese swept into the East Indies, cutting off natural rubber imports from the Far East, capacity goals were raised again, first to 400,000 tons per year, then to 600,000, and finally, in May 1942 to 850,000.

The 850,000-ton capacity goal included 75,000 tons of butyl rubber, 40,000 tons of neoprene, and smaller quantities of Thiokol and butadiene-acrylonitrile rubbers. This too, required a build-up, since 1942 production of neoprene was less than 9,000 long tons, and butyl output was less than 50 million lb.

To get the job done quickly, the

Department of Justice permitted the many plastics, chemical, rubber, and petroleum companies involved to exchange information freely. This the companies did willingly, although they had spent research-and-development money with the hope of retaining proprietary rights to their discoveries. Understandably it was no small task to build up the required styrene capacity, and there was no question of where it would come from: coke-oven benzene, and ethylene from cracking natural gas or refinery streams.

Butadiene, however, was a political problem. The Office of Rubber Reserve wanted butylene as the precursor of butadiene for rubber, while the Office of Petroleum Coordinator wanted butylene for dimerization into a component of aviation fuel. No sooner was this ironed out—by authorizing additional butylene capacity—than the farm bloc in Congress pushed through a

Figure 7–26 A very important material in World War II was the GE-1421 styrene divinyl benzene compound. It provided excellent low-loss properties and would withstand soldering temperatures. (Courtesy, General Electric Company.)

bill demanding that butadiene be made from grain alcohol as well as from petroleum and gas.

Happily all sources were needed. In 1944 almost 500 million lb of butadiene were made from petroleum, but over 700 million lb from alcohol. In that same year alcohol was produced from grain and molasses and by synthesis, in amounts of 313 million gal, 113 million, and 60 million, respectively.

Within a year after the program was authorized, the first among the government-owned plants was producing rubber. Productions reached 182,000 long tons in 1943; 670,000 in 1944; and 719,000 in 1945.

The success of the rubber program was a tribute to the engineering skills of many industries involved and to the effective coordination of the entire effort by various government agencies, especially the War Production Board.

Figure 7–27   After World War II, the market boomed with molded plastics toys and kits. (Courtesy, Revell Incorporated, Venice, California.)

A most important war electronics product was GE-1421 compound (fig. 7–26). This styrene divinyl benzene compound had the remarkable electrical properties of styrene plus the ability to withstand soldering temperatures.

The Arnold Brilhart musical instrument reeds of 1940 and the Finn Magnus all-styrene harmonica of 1944 were important commercial styrene products that stimulated imagination in this field. Impact styrene opened the refrigerator-box market, pushing out the thermosets and steel.

The styrene commercial breakthrough came in 1946 when styrene radio cabinets knocked out the thermosets. The postwar boom in wall tile used a very large volume of styrene. Many cheap and poorly designed toys and housewares hurt the plastics industry in the first postwar years. Toymakers quickly improved their products (fig. 7–27). By 1960 styrene products were to be found in every material market fabricated

Figure 7–28 Photograph cases of molded shellac in the 1860 era (left) gave way to injection-molded styrene cases in 1960 (below).

by every processing procedure. The picture frames of 1860 that had been made of molded shellac were in 1960 made of styrene (fig. 7–28).

Monofilaments and oriented styrene (fig. 7–29) were developed in 1948; light-stabilized formulations in 1949; and heat-resistant varieties in polystyrene foam were invented by Munson in Sweden. Several Dow engineers studied the properties and uses of styrene foam and found methods of manufacturing and commercializing it. The copolymers and terpolymers, styrene acrylonitrile (figs. 7–30 and 7–31), styrene, methacrylate, acrylic butadiene styrene, etc., followed in rapid succession, creating other new markets for plastics. Utilization of the Enthone electroless process for plating on acrylic-butadiene styrene plastics (fig. 7–32) by Marbon put plastics into direct competition with die castings in 1963.

Original phenolic plastics furniture work in 1929 is described in chapter 5. In the years since then,

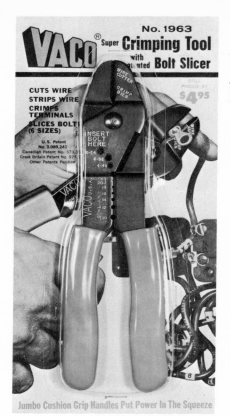

Figure 7–29 Blister packages utilize oriented styrene or acetate sheets thermoformed to the display card.

Figure 7–30 Injection-molded styrene acrylonitrile and an acrylic dial are used for these telephone parts.

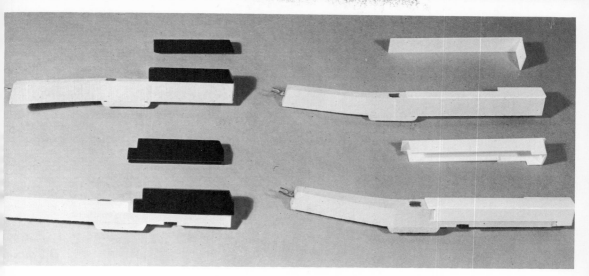

*Figure 7–31   Piano keys moved from ivory, to Celluloid, to acrylic-covered wood.
The all-molded keys of styrene acrylonitrile were developed in the early 1960s.*

*Figure 7–32   The development of plata-
ble ABS materials by Marbon Chemical
Division cut heavily into the use of metals.
The parts shown here are chrome-plated
ABS. (Courtesy, Marbon Chemical Divi-
sion, Borg-Warner Corporation.)*

the laminated plastics came into very large use, particularly for table-top applications. Molders produced drawers for mass-produced furniture. Plastic Industries, Incorporated developed for upholstered furniture a leg of high-impact styrene with a steel through-bolt (fig. 7–33). Polyester decorative overlays were introduced by Syroco and Burwood in the early 1960s. In 1965–66 the use of injection-molded furniture components took off in high volume. In 1966 Plastics Industries, Incorporated, made the first production furniture part of solid high-impact styrene. This was followed by a host of components and, in 1968, an all-plastics chair (fig. 7–34) was introduced. The development of machines for injection molding foamed styrene in the late 1960s has greatly expanded its use for furniture and structural components. Use of beryllium cast molds facilitated very realistic textures and tactile properties.

The famous German chemist,

Figure 7–33 The development of these styrene furniture legs with through-bolt construction started a trend to plastics in furniture in the early 1960s. (Courtesy, Plastic Industries, Incorporated, Athens, Tennessee.)

Figure 7–34 This all-plastics chair was offered in 1968 by Plastics Industries, Incorporated. It was molded of impact styrene. (Courtesy, Plastic Industries, Incorporated, Athens, Tennessee.)

Figure 7–35   Dr. Wallace H. Carothers, shown he
in his laboratory at the Experimental Station, W
mington, Delaware, came to the Du Pont compa
the year after it launched its program of fundament
research in 1927. Dr. Carothers is recognized primar
for his direction of the program of fundamental
search from which came neoprene, the first comme
cially successful general-purpose synthetic rubber, a
nylon, the world's first synthetic fiber comparable
nature's fibers. He died in 1937. (Courtesy, E. I.
Pont de Nemours and Company, Wilmington, De
ware.)

Hermann Staudinger, winner of the 1953 Nobel Prize in chemistry, is widely recognized as the father of modern polymer chemistry. Prior to that time, the organics had been viewed as vague associations of molecules. Staudinger discovered that they are made of large molecules by bonding together small molecules. Dr. Wallace H. Carothers (fig. 7–35), head of Fundamental Research Laboratory at E. I. du Pont de Nemours, advanced Stau-dinger's work, developing neoprene and nylon. Working with Dr. J. W. Hill (fig. 7–36), in 1930 he produced the first synthetic fiber-forming "superpolymer." This forerunner of nylon initiated volume-produc-tion planning. Elmer R. Bolton is credited with the commercial-ization of nylon. Nylon was an-nounced first in 1938 and used as a brush bristle (fig. 7–37). Hosiery filaments were announced in 1939 and offered commercially in 1940.

Figure 7-36   The birth of the first completely synthetic fiber, the forerunner of nylon, is reenacted here. Chemist Julian Hill shows how he [pu]lled molten sample from a test tube, drawing out the thin fiber. (Courtesy, E. I. du Pont de Nemours and Com[pa]ny, Incorporated, Wilmington, Delaware.)

[F]igure 7-37   Comparison photograph of two paintbrushes bristled with tapered nylon. Brush at left has painted 3,500 oil drums; the brush at right is new. (Courtesy, E. I. du Pont de [N]emours and Company, Incorporated, Wilmington, Delaware.)

Figure 7–38 Nylon, which started in Dr. West's toothbrush and in silk hose, found a tremendous market in molded and extruded products.

Nylon came to the molding industry in 1941 (fig. 7–38) and was allocated completely to the war from 1942 to 1944 for parachutes, tire cord, tow ropes and mechanical components. The self-extinguishing potential of nylon enabled it to be used for the Schick shaver. In 1948 nylon qualified for meter gears and took over a large sector of the small-gear production. While nylon has not achieved commodity status, it has become a basic material and is now produced by many plants. The Astroturf pioneered in Houston's Astrodome was developed by Monsanto. It has a nylon face, ribbon-knitted to a 15-oz polyester and nylon backing bonded to a ¾-in. energy-absorbing pad.

Fossil polyethylene was discovered more than 300 years ago. It was produced by nature millions of years before man. The infrared spectrum of this *Fungus subterraneus* or elaterite shows close similarity with industrially produced high-pressure polyethylene.

Polyethylene is credited with winning the Battle of Britain in World War II. Polyethylene-insulated radar cables enabled the limited RAF to knock out the waves of German planes. Samples of this early polyethylene sent to the United States in 1942 for our first radar work facilitated the discovery of the squeeze bottle, as described in chapter 8.

The first systematic investigation of the application of high pressures to polymerization processes was carried out by two Harvard scientists, Professor James Bryant Conant and Professor Percy Bridgeman in the 1920s. The invention of polyethylene resulted from such high-pressure studies started by ICI in 1931.

Many men, M. W. Pervin, J. C. Swallow, E. W. Fawcett, R. O. Gibson, all worked on this program.

Success was achieved in 1936 after numerous failures. Fortunately a leak in the apparatus caused a pressure drop sufficient to permit polymerization of the ethylene and disclose another of our most important plastics. Production facilities planned for the British submarine cable were taken over for radar which functioned on this superior, low-loss, flexible dielectric.

Du Pont and Union Carbide entered into the production of polyethylene in 1941 to meet the United States wartime demands. Dr. George T. Felbeck, Vice President of Carbide and Carbon Chemicals Company, was given the 1949 Hyatt Award for his engineering developments contributing to the large-scale manufacture of polyethylene.

Daniel E. Strain contributed to the polyethylene development at Du Pont. Dr. Karl Ziegler in Germany subsequently developed a catalyst which eliminated the need for

*Figure 7–40   High-density polyethylene proved to be
an excellent packaging material for detergents.*

the high pressures used previously in making polyethylene. Concurrently, Al Clark, J. P. Hagen, and R. L. Banks at Phillips Petroleum Company developed processes for making high-density polyethylene. An alternate process was developed also at Standard Oil Company of Indiana. Polyethylene of ultrahigh molecular weight was subsequently developed by Allied Chemical Company. The Hoola-Hoop (fig. 7–39) craze of 1958 used high-density polyethylene at the rate of 1,000,000 pounds per week at its peak.

The years following World War II saw the squeeze bottle boom into full maturity, taking basic markets from glass and steel. High-density polyethylene proved adequate for the detergent bottles (fig. 7–40). Packaging, closures, wire insulation, paint-brush handles, labware, housewares, toys, and industrial products developed from this amazing material. Rotational molding

Figure 7–41   Links in plastics crystals. The first direct observation of links between the crystals in a polymer was made in 1966 by scientists at Bell Telephone Laboratories. This electron micrograph shows a number of such intercrystalline links, which measure up to 15,000 Å (angstrom units) long and between 30 Å and 300 Å thick. (This is equivalent to about 60 millionths of an inch long and between 0.10 and 1.02 millionths of an inch thick.) H. D. Keith, F. J. Padden, and R. G. Vidimsky crystallized polyethylene from melt after blending it with hydrocarbon, n-$C_{32}H_{66}$, *which is similar to polyethylene but has a shorter molecular chain. When the polymer had cooled, the hydrocarbon —which separated the areas between crystallites—was washed away with solvent. This technique exposed the polymer crystals and the links joining them together. (Courtesy, Bell Telephone Laboratories.)*

developments moved polyethylene, during the 1960s, into many large-product items that were previously untouched by the plastics.

In 1966 studies Bell Telephone Laboratories made the first direct observation of links between the crystals in a polymer, as shown in figure 7–41.

Du Pont reports the discovery of Teflon as "an accident derived from solid research." In his tale, *Three Princes of Serendip,* Horace Wal-

pole called this "serendipity." Early studies of fluorine chemistry led to Freon fluorocarbon refrigerants, and Dr. Roy Plunkett was assigned to produce another type of Freon, using tetrafluoroethylene as an intermediate. Jack Rebok, a laboratory technician, released one of Dr. Plunkett's cylinders of tetrafluoroethylene and found no gas pressure but full normal weight (fig. 7–42). Then the cylinder was cut open and it was found that the gas had be-

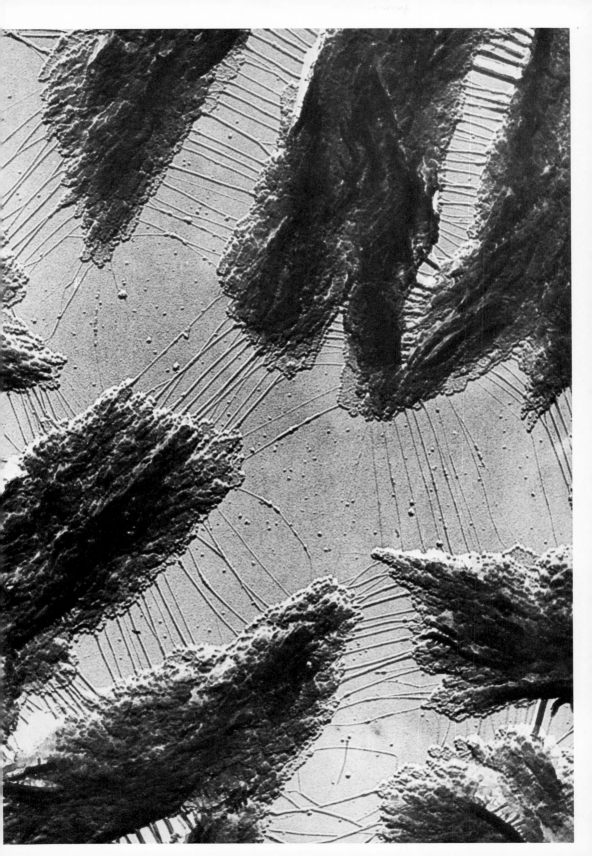

Figure 7–42   Discovery of fluorocarbon polymers in 1938 was made by Dr. Roy Plunkett (right), who holds the original patent. Technician Jack Rebok (left) helped. Chemist Robert McHarness did early fluorocarbon research. In the photograph above, Plunkett and Rebok reenact the discovery at the Jackson Laboratory. (Courtesy, E. I. du Pont de Nemours and Company, Wilmington, Delaware.)

Figures 7–43 Fry pans coated with Teflon fluorocarbon plastics need no grease and will wipe clean after their use. (Courtesy, E. I. du Pont de Nemours and Company, Wilmington, Delaware.)

come a white powder (Teflon). This initial discovery was made in 1938 with pilot production being gained in 1942. This was the first truly higher-temperature organic thermoplastic. This new product of fluorine chemistry filled a great void; it could handle almost every known chemical and did not melt under soldering and fry-pan temperatures (fig. 7–43). Broad and valuable markets were also achieved by its smoothness and antifrictional surfaces. Teflon's electrical properties are uniquely unchanged by temperature and frequency, and it is widely used in radio and power cables. It is not combustible in ordinary atmospheres.

Professor Guilio Natta, 1964 Nobel Prize winner in chemistry, working in 1954 at the Polytechnic Institute in Milan was first to polymerize polypropylene. His distinguished work is of special importance because it proved the potential for making polymer chains of a preconceived pattern, creating another plastics milestone. Polypropylene offered higher temperatures than polyethelene and better freedom from stress cracks, enabling it to be used in some places where other polyolefins were limited. The polyolefins, polyethelene, and polypropylene are among the big three that have achieved commodity status and ubiquitous use. The polyolefins, styrenes, and vinyls comprise 90% of all thermoplastics today.

Urethane plastics introduced in 1955 are borderline between plastics and rubbers. They were developed by research in Germany and in the United States during the period from 1937 to 1954. Mobay Chemical Company was started under joint sponsorship of Monsanto Company and Farbenfabriken Bayer and was closely involved in this development. Many

Figure 7–44   Urethane fibers (Spandex) used in elastic
ski pants caused a boom in ski traffic. These heavy-
duty industrial wheels have a cast urethane elastomer
tread and are extremely wear resistant. (Courtesy,
Hercules, Incorporated, Wilmington, Delaware.)

others worked on the isocyanates and polyhydroxy compounds. The flexible urethane foams were developed in Germany. Development of the foam machine mixing head by P. Poppe and others contributed to the quick acceptance of the foamed urethanes for many products. Polyurethane elastomers were developed in Germany and England during World War II. The polyurethanes in many forms (including an elastic thread, Spandex) have created their own wide and excellent markets (figs. 7–44). Polyester-reinforced urethane, Corfam, was developed by Du Pont in 1958 as a poremeric synthetic leather product for shoes and industrial leather applications. Du Pont reported Corfam to be the result of 200 man-years of research and 3 years of testing. Volume production started in 1965, and the synthetic plastics surpassed the quality of natural leather. Du Pont was forced to close

Figure 7–45   The front case or ogive for the proximity fuze made history with Kel-F. It cured problems previously experienced with ethyl cellulose. (Courtesy, Plax Corporation.)

out the Corfam venture in 1971 after investing more than $100,-000,000; its very high quality was recognized, but the customers wanted low-cost footwear and would not pay the price for a quality product. This is the greatest fiasco that has happened in the plastics industry.

Penton, chlorinated polyether, developed by Hercules Incorporated in 1960 entered the field as a specialty resin for products requir-

ing excellent chemical resistance and good dimensional stability.

Polychlorotrifluoroethylene (CTFE) was developed during World War II by Dr. William Miller of Columbia University for the Manhattan Project. The M. W. Kellogg Division of Pullman Corporation commercialized it under the name of Kel-F in the late 1940s. It was extruded and molded initially by Plax Corporation for gaskets and seals at Oak Ridge. Later it was

Figure 7–46 Vinylidene fluoride, which started as a molding compound, found more important markets in the coating field. This metal siding was given a finish resistant to age and weather with the vinylidene fluoride coating. (Courtesy, Pennsalt Chemicals Company.)

injection molded for the front case (fig. 7–45) of the proximity fuze. Other subsequent markets included self-extinguishing flexible wire insulation, packaging, insulators, and cryogenic seals.

Polyvinylidene fluoride was discovered and patented by Du Pont in the early 1940s and was developed commercially by Pennsalt, who assigned Dr. William Barnhart to this task in the late 1950s. It was introduced commercially as Kynar in 1962. Kynar has developed new markets as a nonburning wire insulation, decorative finish (fig. 7–46) for architectural metals such as aluminum and steel and as a highly chemical resistant material for processing and distributing equipment.

The acetal and polycarbonate plastics are both developed from formaldehyde. The Russian, Butlerov, anticipated such materials in 1859. Einhorn discovered polycar-

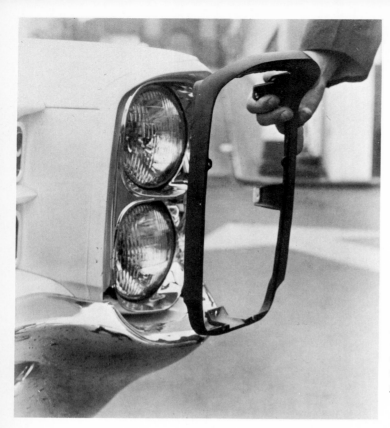

Figure 7-47   Front fender extension
this 1966 model automobile is molded
of Du Pont's Delrin acetal resin. (Cou
tesy, E. I. du Pont de Nemours and C
pany, Wilmington, Delaware.)

Figure 7-48   Polycarbonate resin fou
a large market in window panes and f
lamp globes which could not be broke
by hoodlums. (Courtesy, General Elec
Company.)

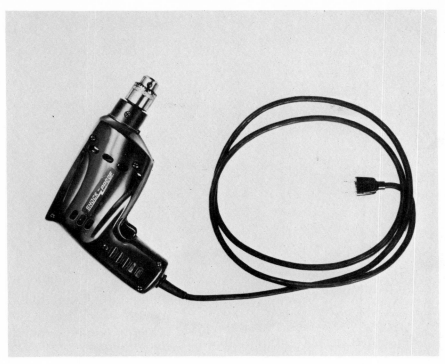

Figure 7–49 Many hand tools used polycarbonate materials to gain double insulation and a shock-free case. Makers of hand tools replaced large quantities of zinc with ABS and polycarbonate in the late 1950s. (Courtesy, Mobay Chemical Company.)

bonate late in the 19th century. Du Pont initiated studies for the production of pure formaldehyde, and Barkdoll's results in 1949 were studied by Robert N. McDonald, which led to a 60-man team that produced Delrin homopolymer acetal resin (fig. 7–47). Du Pont started commercial production in 1959. Celanese later produced their acetal resin, Celcon, a copolymer based on trioxane. Drs. Frank Brown and Frank Bernardinelli developed the Celanese product.

In Germany, H. Schnell, following the formaldehyde trail, reported his polycarbonate resins in 1956, and Bayer started marketing in 1959. In the United States, General Electric Company was also working on this problem and used bisphenol-A for its production of Lexan polycarbonate (fig. 7–48). Mobay Chemical produced their Merlon (fig. 7–49) under Farbenfabriken Bayer license.

Ethylene vinyl acetate copolymers (EVA) were put into pro-

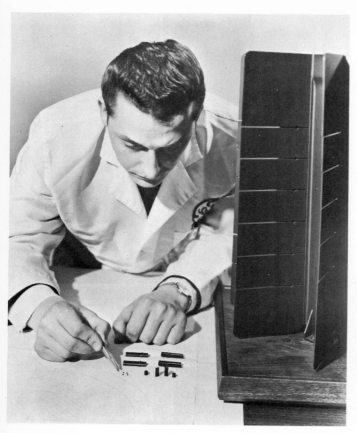

Figure 7–50   The polyimide plastics solved many of the needs of space vehicles for high-temperature-and-vibration-resistant materials. These Vespel parts support and protect the nuclear generato placed on the moon by the astronauts. (Courtesy, E. I. du Pont de Nemours and Company, Wilmington, Delaware.)

duction by Du Pont in 1960. These copolymers offered increased flexibility, elongation, and impact resistance. Du Pont also introduced its Surlin ionomer resin in 1964 providing glasslike transparency, flexibility, and solvent resistance. A polyethylene pipe for gas distribution lines, Aldyl, was added in 1965.

Polyimides introduced by Du Pont in 1962 moved the thermal endurance of the organic thermoplastics up to 750°F and opened numerous other markets (fig. 7–50) in a zone previously limited to the ceramoplastics.

Polymenzimidazoles, developed by Dr. Carl S. Marvel at the University of Arizona in 1965, are providing previously unavailable high-temperature adhesives marketed by Narmco Materials Division for the missile program.

Polyphenylene oxide (PPO), a 1964 development (fig. 7–51) of General Electric Company, offered

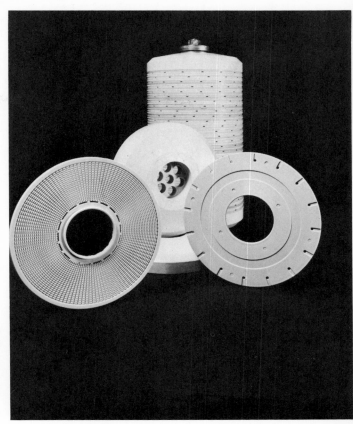

Figure 7–51   Polyphenylene oxide thermoplastics provide autoclavable parts and chemical resistance in this micromembrane filter. (Courtesy, Chemical Department, General Electric Company.)

resistance to hydrolytic attack and stability above the limited temperature ranges of ordinary thermoplastics.

Polysulfone (fig. 7–52), developed by Alfred Farnham and Robert Johnson and introduced by Union Carbide Company in 1965, offered most desirable values for load-bearing structural components at temperatures up to 300°F. It has good arc resistance and replaces some thermosetting materials. The

paraxylenes developed by Dale Pollart and William Gorham were introduced by Union Carbide Company as resins for thin films with high barrier properties. Parylene is used for capacitor dielectrics and for many coating applications.

Noteworthy is the fact that all of the later plastics have been the result of large-group action, much money, and advanced scientific procedures. The cost and complexity of modern chemistry has limited the

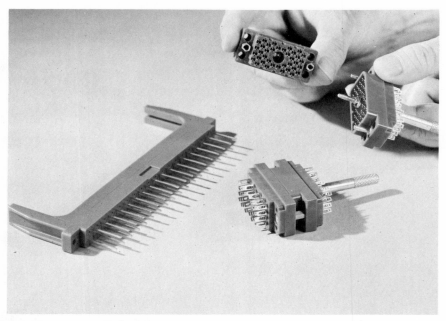

Figure 7–52 Subminiature printed-circuit card-edge connector at left and rack-and-panel cable connector at right are precision molded from polysulfone to gain high-temperature resistance and strength in very thin section. (Courtesy, Union Carbide Corporation.)

potential of lone-wolf scientists like Baekeland and experimenters like Hyatt. Materials of the future are expected to come largely from the great laboratories and large teams, using ultracomplex apparatus and high budgets. We have come a long way from the cellulosics and are snowballing into infinite varieties and alloys of general-purpose and special-purpose materials.

All of the known abundantly available raw materials have been explored and exploited so that there is little likelihood of finding additional basic materials that can be produced at low cost for the high-volume markets. There is room for large growth in the field of alloys, special-purpose materials, and refinement of process in the material-making field. Much work remains to be done on the synthetic inorganic materials. Mold-making and molding-machine developments are expected to make great contribu-

Figure 7–53   The trend to plastics is told very dramatically by this 1970 advertisement of United States Steel Corporation. As F.D.R. said, "When you can't beat them, join them." (Courtesy, United States Steel Corporation.)

tions to future growth of the plastics.

Great strides are being made today against all other competitive materials (fig. 7–53), thus leading up to the Plastics Age in 1983, at which time, it is calculated, the volume of plastics will equal the volume of all the metals and the gasoline business will have become a by-product of plastics.

8

Extrusion and Blow Molding

Plastics Extrusion Evolution*

The present extrusion process emerged near the end of the 18th century when Joseph Brama used the extrusion principle for making lead pipe. The next advance came in the period from 1840 to 1850 when the extrusion process (fig. 8–1) was

*Portions of this chapter were prepared by Jules W. Lindau, III, for *Plastics World,* September 1968.

used to insulate wire with gutta-percha (chap. 1). In 1851 this process was used to insulate the conductors for the cable that was laid between Dover and Calais, which was the first submarine cable.

The first extrusion of insulated wire in the United States was done by A. G. de Wolf. He used an extruder which he developed at the A. G. Day Company in Seymour, Connecticut, in 1858. Thousands of miles of insulated cable were produced, which established the extrusion process for the hand-operated cablemaking industry.

These early extruders, generally called extrusion presses, were rams that were operated manually, mechanically, and hydraulically (fig. 8–2) and which pushed the gutta-percha into a die section through which the copper conductor was drawn. This is the wire-coating process used today with improved feeds, materials, and screws. There were many disadvantages, the greatest being the fact that the process

was discontinuous since it had to be stopped at regular intervals to refill the cylinder. In overcoming this serious limitation, the single-screw extruder was adopted. Ram extrusion was not abandoned with the development of the screw extruder. Many sophisticated developments are being carried on today with the ram extruder, some in combination with screws.

The change from the ram machines to screw extrusion enabled the entire process to advance sig-

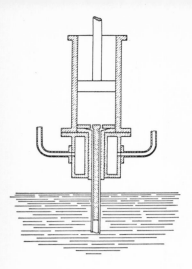

Figure 8–1   In 1845 H. Bewley made plastics pipe in a piston extruder. Gutta-percha and shellac were extruded by this process (Courtesy, Kunststoffe, vol. 55.)

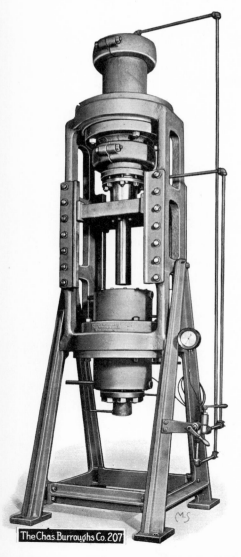

Figure 8–2   This vertical extruding machine is typical of the extrusion presses used by Hyatt and others until the screw extrusion systems for plastics were developed. (Courtesy, The Charles Burroughs Company, Newark, New Jersey.)

Figure 8–3   John Royle developed his screw extruders for rubber in 1879, and today his company is a leading builder of extrusion machines for rubber and plastics. (Courtesy, John Royle and Sons.)

nificantly. The first patent on an extruder employing an Archimedean screw was taken out in 1879 by an Englishman named Gray. John Royle developed a screw machine (fig. 8–3) in the United States in 1880. In 1892 Paul Troester developed a successful screw extruder in Germany, and by 1912 he is reported to have delivered over 500 of these units to the United States for use in rubber and other mixing applications. These early extruders were used to make hose and tire tubes. Apart from gutta-percha and shellac, the first thermoplastic material to be extruded was cellulose nitrate. This material was processed by Hyatt sometime between 1875 and 1880 by using the solvent cold-coat process in a ram extruder. This process is still used for cellulose nitrate.

Many materials were tried with varying results during the period from 1920 to 1936. There was no great success in the extrusion of the styrene and vinyl materials until after the hot-dry extrusion of cellulose acetate became a commercial reality. Prior to that, experimental materials were fed into extruders which were inadequate due to the shortness of their screws and the fact that no provision was made for maintaining heat in the barrel of the equipment. Feeding was performed by the use of hot-strip stock, which was most inconvenient.

An interesting picture of the early extrusion work is reported by P. Willard Crane (Vice President of the Cincinnati Milling Machine Company) for the period from 1928 to 1933. At that time he was working with Ben Fields on the safety-glass interlayer development for Du Pont. The initial studies were made with Celluloid and the early extrusion work used a "stuffer" (the originals were modified sausage stuffers) having adjustable orifice blades mounted on a cellulose block press (fig. 2–3). Later a piston extrusion press was added that could be filled with a low viscosity cellulose nitrate solution which pushed the strip of cellulose into a cold liquid for hardening and removal of solvent. This was called the solvent cold-coat process. Nitrogen pressure was then used to get uniform extrusion pressure. This equipment was found to perform even better with cellulose acetate when it became available in 1931. The success of these studies led to an investigation of the screw extruders.

Up to 1932, no screw extruders had been used for the plastics. The rubber extruders knew very little about plastics or the design of extruders for low viscosity materials. It was then necessary to develop the fundamentals of pitch and depth of flights. Uniform flights were compared with those of decreasing pitch and varying depth. Several experimental screws were cut and recut before satisfactory extrusion was obtained. These fundamental studies proved that the screw extruder was best for pumping and metering plastics with a wide viscosity range. This work led to the extrusion of satisfactory safety-glass interlayers of Celluloid, cellulose acetate, and polyvinyl butyral. Later studies at Du Pont by Bernhardt, McKelvey, Mohr, Fields, and others extended this fundamental extrusion technology. Their studies in the late 1940s and early 1950s made possible a scientific approach to extruder design. Ben Fields developed multiple-ram extruders for the continuous extrusion of cellulose acetate. During the period (1943 to 1950), Ben Fields did basic research and development work on Teflon extrusion, first with dry powder in flighted barrel with shallow screw and secondly with ram and dry powder and then with ram and lubricated TFE.

One of the earliest commercial thermoplastic extrusion machines was designed by Paul Troester in Germany. At about the same time Francis Shaw developed an extruder in England, and not much later the National Rubber Company in Akron, Ohio, also produced equipment that had a longer than normal screw and thermally controlled barrels that were heated by oil, steam, or electrical resistance heaters. The first equipment that was manufactured resembling contemporary machines had a variable-speed drive using a piv gear, a nitrided barrel liner and screw, direct electrical heating of the barrel, air cooling, automatic temperature control, and a screw with a 10-to-1 L/D ratio. Jack Gould and G. Hendrie, Sr., of Detroit Macoid Company developed proprietary methods for the dry extrusion of cellulose acetate

Figure 8–4   Jack Gould and G. Hendrie, Sr., working at Detroit Macoid, contributed greatly to the profile extrusion of plastics in the late 1930s. They developed methods for cellulose acetate that enabled them to use National Rubber extruders as shown here. (Courtesy, Detroit Macoid.)

with rubber extruders and initiated custom profile extrusion, as shown in figure 8–4.

Engineers in many locations were working then on the hot-melt extrusion of thermoplastics. There was no general source of information; machinemakers and extruders had done no experimental work on the plastics and knew very little about their own products. James T. Bailey at Plax Corporation made extensive studies of extrusion and heat plasticization in preparation for the hot-melt blow molder. Initial studies were designed to find methods for the dry extrusion of cellulose acetate as shown in figure 8–5. These experiments led him into further extrusion studies that developed large diameter rod-and-slug extrusion, biaxially oriented film and sheet, slipper-plate dies, orienting and winding, monofilaments, tube winding, the blow extrusion of layflat tubing, and blow molding.

In the early 1940s, styrene sheets

Figure 8–5 Shown here is an early experimental extrusion machine as developed by Jim Bailey at Plax Corporation in 1938. (Courtesy, Plax Corporation.)

were compression molded. Wood strips laid at the edges of the platens confined the material and compressed as the material became densified. The biaxial orientation process (fig. 8–6) eliminated the uneconomical compression process for styrene sheet. Thin oriented styrene sheets were also compression laminated into very strong thick sheets.

Heavy-wall tubing was extruded on a mandrel at Plax with excellent density and control as shown in figure 8–7. Large-diameter styrene tubing was produced by winding the hot extrudate on a mandrel with a lap joint as shown in figure 8–8. Cylinders made by this process were used early in World War II for high-frequency coil forms, radome, and radar components.

The 36-inch sheet machine of 1941 (fig. 8–9) was first to extrude the plastics sheet or film with biaxial stretch, which also eliminated die marks, gaining completely clear surfaces. This led to the development of oriented styrene, which pro-

Figure 8–6   Initial work on the biaxial stretching of cellulose acetate sheets at Plax is shown here as done in 1941. (Courtesy, Plax Corporation.)

Figure 8–7   Shown here is the first direct extrusion of heavy-wall tubing at Plax Corporation. (Courtesy, Plax Corporation.)

Figure 8–8   The extrusion of heavy-wall, large-diameter tubing was achieved initially at Plax by winding the molten extrudate on a mandrel with pressure applied to bond the lap joint thus produced. (Courtesy, Plax Corporation.)

Figure 8–9   Developmental work on oriented styrene was done on this early machine of Plax Corporation. Grippers grasped the ribbon of hot plastics as it exited from the extruder and stretched it in width as it was also stretched in length. Shown at the upper left is the original Strauch machine for automatic thermoforming of sheets. (Courtesy, Plax Corporation.)

duced a high-strength thin styrene sheet. Polyflex oriented styrene is used extensively for packaging, table mats, windows for envelopes, and thermoformed products. The early radome shown in figure 8–10 was made in 1943 for Massachusetts Institute of Technology from preshaped pieces of oriented styrene, subsequently bonded in a compression mold by heating under pressure. His work in blow molding has tended to overshadow his numerous contributions to the extrusion of the plastics.

Saran monofilament was initially extruded in 1943 (fig. 8–11) by Dow Chemical Company. It found many markets for window screen, woven seating, etc. Later extrusion developments opened up the film market for Saran vinylidene chloride where it made a tremendous

Figure 8–10   This 1943 radome for Massachusetts Institute of Technology is 30 in. in diameter and 36 in. high. It was compression laminated from thin oriented polystyrene sheets at Plax Corporation. (Courtesy, Plax Corporation.)

impact on the food industry for the packaging of processed foods.

One of the first direct electrically heated extruders was made by the Hartig Engine and Machine Company for the extrusion of nylon in the late forties.

The availability of polyethylene initiated further studies of sheet extrusion which matured in the 1946–47 layflat blown tubing development shown in figures 8–12 and 8–13. This single Plax development

opened up the large market for polyethylene film packaging.

The extrusion coating of paper with polyethylene was developed in 1948 by Charles Fields in Du Pont's Arlington Laboratory. The first production was made by St. Regis Paper Company at Oswego, New York in 1950. The need for a translucent glazing material was envisioned by Harold Warp in 1920. In 1924 he started marketing a product made of cloth filled with french

Figure 8–11   Saran vinylidene chloride monofilament was first extruded in 1943 at Dow Chemical Company.

talc and treated with highly refined petroleum wax and stearate, as depicted in figure 8–14. This was converted to a transparent material in 1938 by the use of cellulose acetate butyrate. Subsequently the vinyls, acrylics, and polyethylenes were used to produce the Flex-O-Glass products as they became available. The experience with outdoor weathering gained from chicken-coop work enabled Harold Wrap and Gordon Kline to establish standards for airplane fabric dopes for World War II.

An important innovation by Jules Lindau at Southern Plastics Company was the extrusion of a 24-in.-wide sheet of methyl methacrylate through an 8-in. tube die for the temporary housings program that followed World War II.

In the late 1950s C. Maillefer, a Swiss engineer, developed a screw (fig. 8–15) that would allow for more efficient and uniform melting

Figure 8–12   Here and in figure 8–13 are shown the original (1947) experimental set-up to prove the feasibility of blown tubing. (Courtesy, Plax Corporation.)

Figure 8–13   Here and in figure 8–12 are shown the original (1947) experimental set-up to prove the feasibility of blown tubing. (Courtesy, Plax Corporation.)

Figure 8–14 Early work with translucent glazing for chicken coops was done here by Harold Warp in 1920. This was converted to a transparent cellulose acetate butyrate product in 1938.

Figure 8–15 The Maillefer or BM screw employs a thin gap as shown at G as an exit for the fully plasticized material. Unplasticized material must remain in the area below until it reaches the essential fluidity to exit via the thin gap.

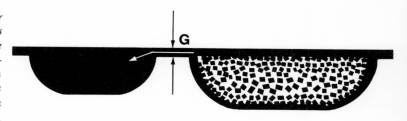

Figure 8–16 The Engel melt extruder divided the functions into separate sections, melting on a rotating plate and pumping with a screw: (a) feeder chute; (b) melting plate; (c) extruder barrel.

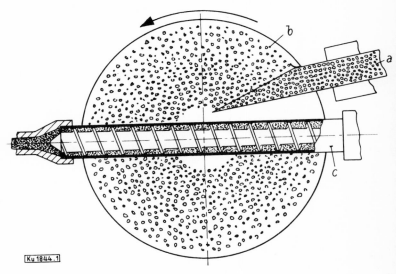

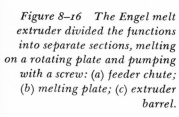

of the polymer by changing the conventional screw to a double-channeled intermeshing screw in which the feed channel conveys the dry granule or powder through a converging area while the metering has a diverging area. As the material is melted, it goes over the land of the intermeshing flight and feeds into the diverging area. As the solid is moved forward, there is less and less space available. The melt continues to flow over the flight into the diverging section, which continues to enlarge. This system improved the transfer of heat from the barrel to the material, resulting in more uniform melt or extrudate.

Another solution to this problem was found by T. Engel, a German engineer, during the same period. He developed a method (fig. 8–16) of dividing the two functions of an extruder into separate sections, one,

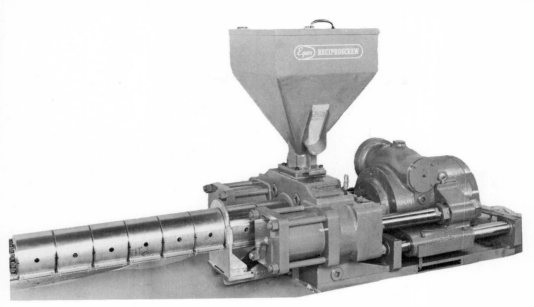

Figure 8-17 The reciprocating screw serves to plasticize the charge and inject the molding material. This 1952 development greatly expanded the injection-molding process for thermosets and thermoplastics. (Courtesy, Frank W. Egan & Company, Somerville, New Jersey.)

the melting of the polymer, which was done on a flat rotating plate that was heated, and, the second, the pumping of the melt in the usual metering section.

The most important extrusion development of the sixties was the near universal adoption of the reciprocating screw plasticizer and injector (fig. 8–17) invented in 1952 and patented in 1956 by William H. Willert. This gave the injection machine user the first real improvement since the invention of the Gastrow torpedo of 1932 as described in chapter 6. This development also expanded the thermosetting field again by the use of the reciprocating screw for automatic transfer and injection molding of thermosets as described in chapter 4.

The Transfermix extruder introduced in 1968 by Sterling Extruder

Figure 8–18 The Transfermix extruder introduced by Sterling Ex-truder Corporation in 1968 was designed to improve and combine the functions of intensive mixers and extruders.

Corporation (fig. 8–18) combined the functions of intensive mixers and extruders. It was developed by Dr. M. S. Frenkel in England and by Uniroyal in the United States.

During the 1960s the development of the theory and practice of extrusion in polymer processing was fairly slow as compared to the progress in the late 1940s and early 1950s. During that earlier period, one advance followed another in rapid succession and led to the de-

velopment of methods for calculating the performance of extrusion screws for plastics. The equations which allow the calculation of screw extrusion flow problems were summarized by H. R. Jacobi in a text entitled *Screw Extrusion of Plastics: Fundamentals, Theory* (Gordon and Breach, New York, 1963). Such calculations have become an accepted tool for the processing engineer in predicting the performance of extruders, and much experimental work has been eliminated.

To consider more of the developments in extrusion procedures, it is necessary to recall the twin-screw extruder, which dates back almost as far as the original single-screw unit and which has been continuously improved. The twin-screw extruder with counterrotating screws provides good temperature control and is used to extrude rigid PVC for pipe. An additional development has been the construction of a two-stage twin-screw extruder which actually uses four screws, the first two used to melt and combine the materials which then feed into a second stage containing two more screws that are used for additional intensive mixing and control of uniform flow. This machine is manufactured by Anger. There are a number of twin-screw extruder manufacturers and a wide range of designs.

In the late 1950s Maxwell and Scolora at the Princeton Plastics

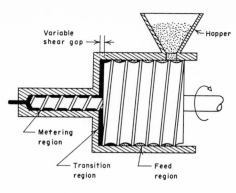

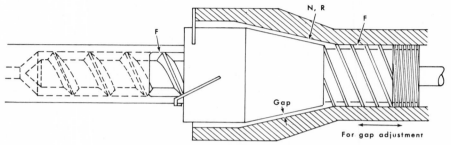

Variable
shear gap →

← Hopper

Metering
region

Transition
region

Feed
region

Figure 8–19   The elastic melt extruder as built by KPT Manufacturing Company.

N, R

F

F

Gap

For gap adjustment

Figure 8–20   The Werner and Pfleider Plasticator single-screw extruder makes use of a feed screw to force the polymer to flow through the gap between the cone and its mating surface inside the barrel. The melted polymer is then pumped out by the screw.

Laboratory developed what has been termed an elastic-melt extruder. This machine is shown schematically in figure 8–19. It was the first continuous-flow extruder to be developed. The elastic properties of the material are used rather than the viscous properties which cause pressure to be developed in the material while it is being sheared between a rotating plate and a stationary plate. The principle that this type of extruder operates on was de-

scribed by K. Weisberg in 1947. The addition of a short screw to the rotating disc by K P T Manufacturing Company is a step beyond the Maxwell and Miner design. The addition of the screw makes it possible to process under conditions in which there is a fair degree of back pressure.

A most significant development in the late 1960s permitted the removal of volatiles within the single-screw extrusion system. Venting

Figure 8–21   This Aragon front-end drive extruder has the screw protruding into the hopper. The drive is at the metering end of the screw. (Courtesy, Aragon Products Incorporated.)

improved the quality of the extrudate and the production rate. This was accomplished by using a two-stage screw and venting either to atmosphere or by attaching a vacuum to one vent to draw off volatiles and remaining water vapor.

Although for most thermoplastic extrusion, the ram extruder has been superseded by the screw extruder, alternate designs are reported to be under study. The Werner and Pfleiderer Plasticator single-screw

extruder (fig. 8–20) combines several features. A feed screw forces the material to flow between the cone and its adjacent surface inside the barrel. The melted material forms into small cylinders upon reaching the enlarged gap and then enters the feed hopper of the screw. Volatiles escape at this point. The screw then pumps the material into the die. A unique development of the late 1960s is the Aragon front-end drive extruder (fig. 8–21) pro-

Figure 8–22 Extruders today are very complex and sophisticated machines, as illustrated by this Egan 10″ × 12″, 10:1 enclosed-screw, melt-fed extruder with underwater pelletizer. (Courtesy, Egan Machinery Company, Somerville, New Jersey.)

duced by Aragon Products, Incorporated. This unit has the screw protruding into the hopper where a minimum-root diameter improves hopper feed. The drive is located at the metering end of the screw where the stress is the greatest.

It is recognized that new materials are being developed constantly. In order to use these materials under the best conditions for retention of properties and economical processing, new methods must be developed to handle them.

Extrusion is the most versatile method of manufacturing at the present time, being incorporated with such applications as blow molding, injection molding, automatic-transfer molding of thermosetting materials; and the process of manufacturing coated webs, profile shapes, sheets, rods, and tubes. The new polymers require more sophisticated extruders (fig. 8–22), and the design thereof is made by computers instead of by experiment. Die design is well understood at this

Figure 8–23 The size of extrusion dies appears to be unlimited, as shown here. This 130-in.-wide die can produce sheet up to ½-in. thick with an output capacity of 2,000 lb per hr. (Courtesy, G. M. Plastics Company, Granby, Quebec, Canada.)

time, and extrusion dies of tremendous size may be used (fig. 8–23).

The first reported thermoset extrusion was demonstrated to Condensite Company in 1912 by Harry P. Taylor. He extruded 3-in.-diameter rods 4 ft long but did nothing with his work. The next recorded work on thermosetting extrusion was done successfully by Gus Sheafer and Henry Holmgren at Plastics Engineering Company in 1944. They used a ram extruder with high-frequency preheat to extrude a cloth-filled phenolic material into bazooka barrels 20 ft long, 5 in. in diameter, and with a ¼-in. wall. Badger Plastics Company extruded phenolic clarinet body tubing in 1947. This was followed by work in Germany and at York Industrial Plastics, licensed under developments of Sud-West-Chemie, New Ulm, West Germany. York Industrial Plastics, now a division of The Budd Company, has developed a proprietary line of rods and tubes plus many structural shapes.

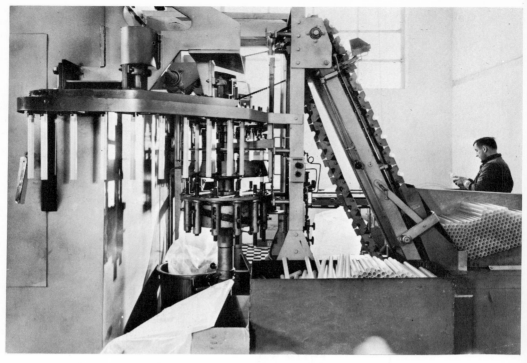

Figure 8–24   The preformed parison or cold-parison process is a modern version of Hyatt's early blow-molding process as depicted in figure 2–5. Precut sections of tubing are heated and blown in this process. (Courtesy, Tuboplast-France, Paris, France.)

All thermosetting materials can be extruded today.

Blow Molding

The early blow forming of Celluloid by Hyatt described in chapter 2 was in essence a thermoforming operation, and this early process is used today in the preformed parison or cold-parison process blow-molding machine (fig. 8–24). Hyatt's machine is shown in figure 2–5. Precut sections of tubing were heated and expanded by steam pressure while the tube was clamped in the mold as depicted in figure 8–25. A wide variety of toys, baby rattles, and novelties were blow formed by Hyatt.

Hartford Empire Company (Emhart Manufacturing Company), a developer and manufacturer of glass machinery, envisioned the oncoming competition from plastics and initiated its Plastics Experimental Station in 1933. This

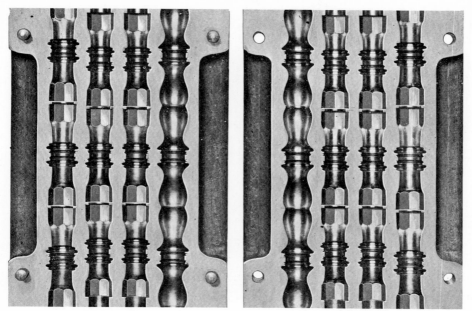

Figure 8–25 The Hyatt internal-pressure blow-forming process made use of celluloid tubes, heated to the plastic state and inserted in a mold such as this, where the tube was expanded by internal steam pressure. This process (also called the cold-parison process) is used today in the preformed or cold-parison blow-molding machine as shown in figure 8–24. (Courtesy, The Charles Burroughs Company, Newark, New Jersey.)

was incorporated as Plax Corporation in 1935. Fernplas Corporation was formed in the early 1930s by E. T. Ferngren and W. H. Kopitke, who developed several bottlemaking methods (fig. 8–26) with patented procedures (see chapter 6). W. H. Kopitke invented the injection blow-molding process (fig. 8–27) in 1938 with patents being issued in 1943. Plax acquired a license under these patents and ultimately acquired the men and the company.

James T. Bailey, a glass consultant, joined Plax in 1937 to expand the development of the hot-melt extrusion blow-molding process. Bailey made an exhaustive study of extrusion in preparation for the blow-molding experimental work. This led to the development of the extrude-and-blow machine.

The first Plax machine (fig. 8–28) had a single crosshead, directly under which was located a one-cavity blow mold. Tubing was extruded downward to a proper

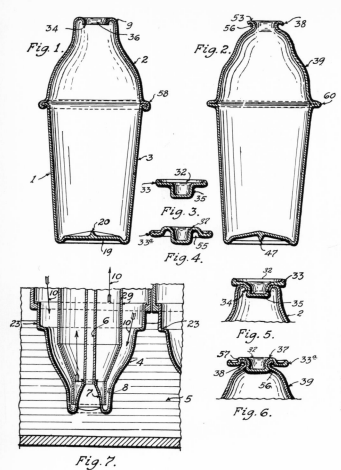

Fig. 1.

Fig. 2.

Fig. 3.

Fig. 4.

Fig. 5.

Fig. 6.

Fig. 7.

Figure 8–26   Fernplas Corporation was formed to explore the plastics bottle potential. Shown here is their 1930 approach. Noteworthy is the fact that bottles of this type are being made today by the thermoforming process, as shown in figure 6–44.

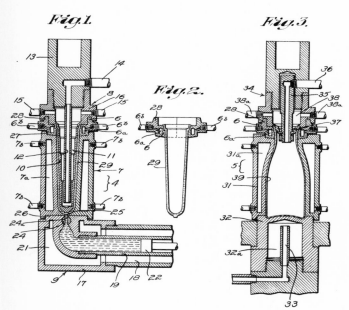

Fig. 1.

Fig. 2.

Fig. 3.

Figure 8–27   W. H. Kopitke of Fernplas Corporation was first to develop the injection blow-molding process as illustrated by this 1938 work. The machine is shown in figure 8–28.

Figure 8–28   This is the first hot-melt extrusion blow-molding machine as developed by Plax Corporation in 1938.

length after which the extruder was stopped and the blow mold closed. In closing, the mold halves pinched the bottom end of the tubing, and air was introduced through the core tube of the crosshead, forming the bottle. After cooling, the mold was opened, the blown product removed, and a secondary trimming operation performed. This initial machine was modified to handle four molds in sequence as shown in figure 8–29 and became the first production blow molder. Just prior to our entry into World War II, this machine produced huge quantities of acetate Christmas tree balls, replacing the glass ornaments that had previously come from Europe. Several million styrene urinal specimen bottles were blown for the medical corps during the war. During this period, acetate bottles were blow molded in large volume for the oil industry's field-sampling program. Styrene bottles for the air

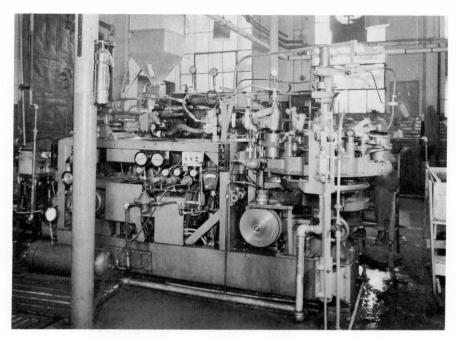

Figure 8–29   The first production blow-molding machine at Plax Corporation. It used the bobbing extruder in the extrusion blow process.

transport of battery acid were supplied during the Korean War.

Owens Illinois Glass Company initiated their plastics studies in 1935 with work being done by Stephen T. Moreland, Victor E. Hoffman, and Parker C. Tracy. The S. T. Moreland patent (fig. 8–30) on an injection blow process was filed in 1939. His mechanism was mounted on a 1940 8-oz Lester injection molding machine and produced some bottles for the U. S.

Army Medical Corps in World War II. Their program was shelved when no postwar uses for blow-molded cellulose acetate or polystyrene were envisioned. Polyethylene had not been used at that time.

In December 1942, the first polyethylene bottle was blown. Fantastic as it may seem today, this lusty infant was a disappointment to its makers. The lack of transparency and the rigidity of this first squeeze bottle caused it to be discarded as

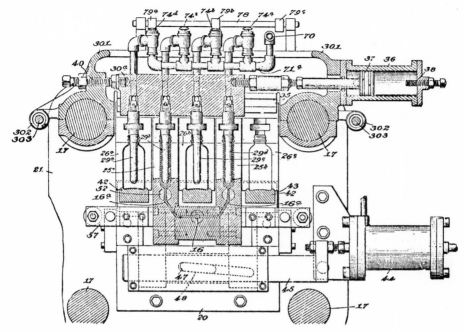

Figure 8–30 Owens Illinois Glass Company filed the Moreland patents in 1939 on an injection blow process that ran in an 8-oz Lester injection machine.

another useless curiosity. The goal then was an unbreakable bottle to replace glass and this particular creation was the result of war's need for better radar insulating materials and engineers' curiosity.

In December 1942, 200 lb of the newly developed polyethylene material was flown to the United States from England under an experimental contract of the government to extrude ⅛-in.-diameter spaghetti radar insulation. Excitement gener-

ated by the ease of extrusion of this unique new material prompted the responsible engineers to collect the scrap and try it in the blow molder. While it appeared to be an ideal blowing material, its potential as a functional dispensing bottle was overlooked until 1944.

Present at the birth of the polyethylene bottle in 1942 were James T. Bailey, William Kopitke, Brian Minalga, Ray Winchester, Gerard C. Heldrich, and R. S. Jessionowski.

Figure 8–31   James T. Bailey led the Plax extrusion and blow-molding developmental work. He conceived ideas, proved them mathematically, and then built crude mockup machines in his basement workshop to prove function.

James T. Bailey (fig. 8–31) was given the Hyatt Award in 1951 for his work in perfecting the hot-melt automatic blow-molding machine and for his important extrusion developments. He was also given the famous Edward Longstreth Medal from the Franklin Institute in Philadelphia for his plastics work. During the intensive research and development period, Bailey was assisted by L. H. Fehrenbach, H. A. Pratt, R. W. Canfield, R. S. Jessionowski, J. Wilkalis, Ed Lorenz, and

Karl Peiler. Mr. C. Paul Fortner was selected from Du Pont to carry on Bailey's work after his retirement.

Samples of the polyethylene bottle were given to the Plax salesmen in 1944 for exposure to potential customers, and Hudson Talcum Powder Company was the first user of the new squeeze bottle. They used a wooden plug with a central hole to produce a powder spray from an 8-oz Boston round bottle. Jules Montenier of Stopette was

Figure 8–32 Jules Mon-
tenier's Stopette spray bottle
was the first high-volume
squeeze bottle and did much
to develop the blow-molding
market.

soon to follow with his underarm deodorant package (fig. 8–32), and this gave the squeeze bottle national publicity. Spray-Net followed immediately with a spray-plug atomizer and at this point, the polyethylene spray bottle took off in a real explosion with customers fighting for bottles.

The first bottles offered to the trade were replicas of the standard Boston round in 2-, 4-, 6-, and 8-oz sizes. Shown in figure 8–33 is the first custom-styled plastics blow-molded package as designed by Walter T. Heintze. Its appearance on the market led to an avalanche of custom-styled bottles such as the tomato catsup bottle (fig. 8–34), which opened the food market and set another high in plastics products sales records. Shown in figure 8–35 is the first captive-cap closure on a 4-oz Boston round bottle as invented by Wayne F. Robb at Shaw Insulator Company. This was the first use of the polyethylene hinge.

Other special interest milestones

Figure 8–33 This Anahist bottle was the first custom-designed blow-molded package, styled by Walter T. Heintze in 1947. All previous packages had used the common Boston round. This bottle started an avalanche of custom-styled packages.

Figure 8–34 This catsup dispenser formed as a red tomato with green closure was the first squeeze-bottle food package. It was designed by Morris Friedman and did much to popularize the polyethylene bottle.

Figure 8–35 This is the first molded hinge product as designed by Wayne F. Robb for Shaw Insulator Company in 1945. Many of the early spray devices and closures were designed by Wayne F. Robb, Robert Musel, and Don Biklen.

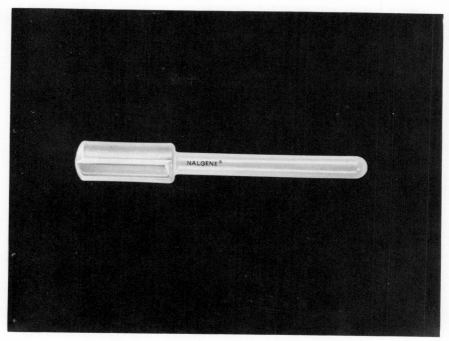

Figure 8–36   Shown here is one of the smallest blow-molded products. It is a one-piece medicine dropper designed by Fred W. John at the Nalge Company, Incorporated, Rochester, New York.

in blow-molded products are the medicine dropper (fig. 8–36), the bellows (fig. 8–37), and the styrene drug bottle (fig. 8–38).

R. W. Canfield was responsible for the Plax BNR machine (fig. 8–39) developed in 1950 for cold cream jars and other such products. It used the basic injection-blow process.

The lining development was the most important secondary contribution to the plastics bottle pro-

gram. This permitted the polyethylene bottle to be used for packaging materials that could not be contained in unlined bottles. The lining materials and procedures were developed first by Jules Pinsky and Alvin Nielsen at Plax Corporation. The first commercial product to be packaged by this process was a Wildroot hair preparation in 1955. Another important contribution was the Kriedel flame-treating process to facilitate printing; corona-

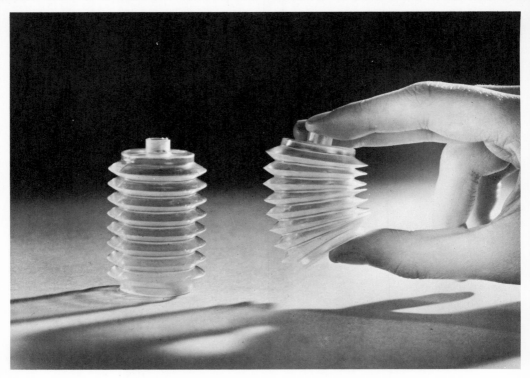

Figure 8–37   Ethylene vinyl acetate is blow molded to form this bellows, which gives voice to a doll.

Figure 8–38   These styrene bottles were developed for the packaging of food and drug products. They were injection blow molded by Wheaton Plastics Company.

Figure 8–39   The Plax BNR machine of 1950 was of the injection blow type for producing wide-mouthed bottles and cold-cream jars.

discharge systems were also developed for some applications.

Elmer E. Mills, a prominent Chicago injection molder, entered the field in 1949 with a unique process which was ultimately taken over by Continental Can Company. Mills used a conventional extruder to deliver a continuous tube of plastic material which indexed with molds as they moved into position on a continuously rotating table, as shown in figure 8–40. The mold halves clamp when a full length tube is in place and then move into a blow position for inflation by a needle. As the mold continues its travel, it comes into a cooling area and then to the opening position with air blast ejection.

C. C. Coates of Royal Manufacturing Company developed a third process in 1951 illustrated in figure 8–41. This is similar to the Mills machine except that the table is vertical. Kautex machines (fig. 8–

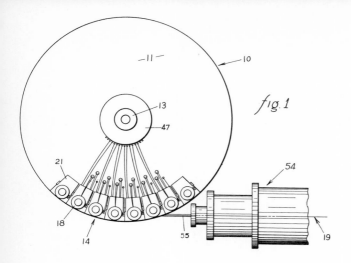

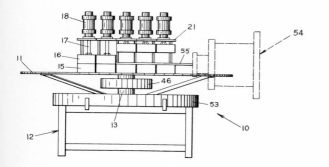

Figure 8–40   The Elmer E. Mills 1949 bottle-making process extruded a continuous tube with the blow molds clamping in sequence on the hot extrudate for inflation by a needle.

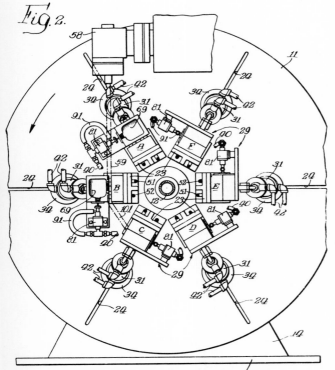

Figure 8–41   The Coates process of 1951 was similar to the Mills invention except that extrusion and mold rotation were in a vertical axis.

Figure 8–42   The Kautex Model B 1 blow-molding machine. The Kautex machines came into this country in the early 1950s. They used the extrusion blow process with a rising mold.

42) were then coming in from Germany; they extruded a tube in a continuous downward motion and indexed on a blow tube located below the extrusion position of the mold halves. When the plastic tube reached the essential length, the mold halves closed, and a knife cut the oncoming tube permitting the mold to drop to blow position where it did not obstruct the oncoming tube flowing continuously from the extruder.

Notwithstanding the earlier contributions of Plax Corporation and Owens Illinois Glass Company, injection blow molding did not become a commercially feasible process until the middle of the 1950s, when Wheaton Plastics Company developed its first proprietary machine for high-volume production. In early 1950, the Wheaton Glass Company initiated its development efforts in the field of injection blow molding, and in the beginning

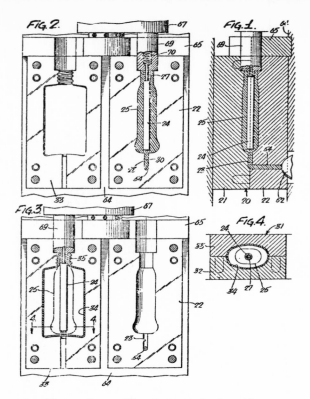

*Figure 8–43 The Borer injection blow
process of 1953 was designed for operation in an injection-molding machine.*

worked under patent rights (fig. 8–
43) granted to it by the Swiss inventor Alfred Borer. Through these
pioneer efforts and the subsequent
development work carried on at
Wheaton Plastics Company, the injection blow-molding process has
evolved as a significant factor in
supplying the market demand for
containers in sizes under one quart.
The invention of the reciprocating
screw made a major contribution to
the injection blow-molding process.

The advent of the blow-molded

bottle stimulated many two-piece
bottle developments by the injection molders. The cubic bottle (fig.
8–44) was made by Shaw Insulator
Company in 1949 for the Medical
Corps. The bottom was crimped
metal as used also for metal cans.
Boonton Molding Company introduced an injection-molded collapsible tube for Ciba. Injection Molding Company, later Imco Container
Company, was attracted to the bottle business by its success in the
cosmetic industry and entered the

Figure 8–44 The appearance of blow-molded poly-ethylene bottles stimulated many injection-molded packages. The cubic bottle here was designed by Shaw Insulator Company for the Medical Corps. It has a crimped metal bottom like that used for tin cans. The Mennen bottle was made in two pieces with the bottom welded in place by Imco Container Company.

industry in the late 1940s with a two-piece injection molded bottle as shown in figure 8–44. They soon changed over to a blow-molding machine which used a multiple-orifice manifold. Each orifice was valved to provide sequential delivery of a parison into a blow mold. A machine of this type was also marketed then by Kato Seisakusho Company and Moslo Machinery Company.

Owens Illinois Incorporated, who had done original work as de-scribed previously and had worked closely with Plax at one time, started production use of a machine having a movable neck ring that was filled by the extruder and then moved downward followed by the extrusion into a position for clamping in the mold. This machine also provided good neck and lip finish.

All of the original blow molders except Wheaton Plastics Company and Owens Illinois have lost their identity. Plax was absorbed by Monsanto Company after a brief period with Owens Illinois Incorporated, Royal went to Celanese Plastics Company and then to American Can Company. Imco Container Company was taken over by Rexall Drug and Chemical Company and later by Ethyl Corporation. Elmer E. Mills Company was taken over by Continental Can Company. Union Carbide Company, Phillips Products Company, and Foster Grant Company developed blow-molding facilities as one means to expand the use of their molding materials. The last decade has seen many captive and custom blow-molding facilities developed as commercial machines became readily available.

Many machine builders in the United States, Japan, Germany, and Italy jumped into the field in the late 1950s with a variety of general-purpose and custom-built blow-molding machines. Most injection machinery builders now offer blow-

Figure 8–45   High-density polyethylene solved the packaging problem for the bleaches and detergents and tremendously expanded the market for blow-molded bottles.

Figure 8–46   This new experimental Coca Cola bottle blow molded by Monsanto Company is the forerunner of an avalanche of no-return bottles coming up in the 1970s.

*Figure 8–47   This plastics fuel tank was the first poly-
ethylene high-density blow-molded unit. Produced in
1966 by Phillips Products Company, it weighed 9½ lb
and held 20 gal.*

molding machines also. The development of linear polyethylene
permitted its use for packaging
bleaches and detergents and established another milestone for blow
molding. The tremendous bleach
market was met with greatly increased plant capacity to produce
the half-gallon and gallon jugs as
depicted in figure 8–45. Some producers installed their own machines
to ensure adequate capacity and
cost control.

The larger blow molders of the
1960s designed and built their own
machines. Secrecy was the order of
the day, and all design and opera-
tional details were guarded very
closely because of the highly competitive markets that they served.
Controlled-parison extrusion, dual-
plane extrusion blowing, and other
refinements of the 1960s have added
greatly to the versatility of the blow-
molding process. The last decade
saw the blow molding of vinyl

Figure 8–48 This "Giant" blow-molding machine of Modern Plastics Machinery Company is representative of many such developments in the 1960s.

resins start to mature, and they are expected to take over a large part of the food packaging markets in the 1970s. The first blow-molded Coca Cola bottle is shown in figure 8–46 as produced by Monsanto Company for the market-testing program. Automobile gasoline tanks (fig. 8–47), beer and food containers, machine components, milk bottles, garbage cans, luggage, housings, scientific apparatus, and blow-molded machine components have all developed from the squeeze bottle in its first thirty years of progress. A large-sized blow molder of the late 1960s is shown in figure 8–48. Numerous studies to produce degradable plastics materials are under way to minimize the garbage disposal problem that is being created by the no-return plastics packages that do not disintegrate when buried.

9

Vulcanized Fibre,
Laminated and Reinforced Plastics

Vulcanized fibre was invented by the Englishman Thomas Taylor in 1859. Having made no progress in England, he came here and received American patents in 1871. The Vulcanized Fibre Company, licensed under his patents, was formed in 1873. At that time the infant electrical industry needed sheet insulation material, having only rubber, mica, paper, and cloth for this use. In the making of vulcanized fibre, especially prepared paper is gelatinized in a zinc chloride bath. This product is rolled up on a cylinder under pressure that bonds the plies. When cut from the roll, the resulting build-up is pressed flat and cured by immersion in successively reduced concentrations of zinc chloride that leave the converted cellulose or vulcanized fibre. This is a very tough, easily fabricated, arc-resistant insulating material. Unfortunately, it has high moisture absorption that causes decreased electrical values and warpage. Vulcanized fibre continues to be an important material, but it has lost many of its original electrical applications to the laminated phenolics. Its toughness, formability, arc resistance, and low cost let this old product continue as a basic material.

Other pioneer fibremakers include The Kartavert Company in 1876, Delaware Hard Fibre Company in 1892, American Hard Fibre Company in 1894, and Diamond State Fibre Company in 1895. All were absorbed in 1901 by American Vulcanized Fibre Company except Delaware Hard Fibre Company and Diamond State Fibre Company. Continental Fibre Company was formed in 1905, and they purchased Diamond State Fibre Company in 1929, forming Continental Diamond Fibre Company, which was taken over by the Budd Company in 1955. Iten Fibre Company was formed in 1918. National Vulcanized Fibre Company was established in 1922.

Spaulding Fibre Company was started in 1873 to press scrap leather into an "artificial leather sheet" for shoe components. They entered the vulcanized fibre field in 1906.

One of the early natural molding resins was used also as a binder for sheet plastics. The shellac laminates were developed in Switzerland by Emil Hoefely who brought them here in 1908, licensing the electrical industry to make these early shellac laminates. The shellac laminates were extensively used for many years as transformer insulation. Until 1950, paper bonded with shellac was manufactured by General Electric Company and sold as Herkolite insulation.

The advent of Dr. Baekeland's phenolic resin as a binder for the laminated plastics sheets, rods, and tubes was a boon to the electrical insulation industry since it did not absorb much moisture, was quite stable dimensionally, and had good wet dielectric strength. Dr. Baekeland started this work with Westinghouse, which spawned Formica Insulation Company, the first commercial producer of laminated plastics. General Electric Company, Richardson Company, Mica Insulation Company, and Panelyte Corporation followed along with the makers of vulcanized fibre into this field. Spaulding Fibre Company started making laminated phenolic in 1923 and Diamond State Fibre Company in 1914. R. R. Titus left Diamond State Fibre Company in 1928 to form Synthane Corporation, concentrating on the phenolics and fabricated parts. John Taylor started Taylor Fibre Company in 1932 to produce fibre and the laminated plastics. These two companies were merged in 1968 to form Synthane-Taylor Division of

Alco Standard Corporation.

Dr. C. E. Skinner, a Westinghouse engineer who was working with the shellac laminates, reports his introduction to the phenolic laminates as follows:

I read Dr. Baekeland's Franklin Institute paper about his new resin, a condensation product of phenol and formaldehyde. The characteristics of the new material as given in the paper immediately appealed to me as something worth investigating. At once I asked the Purchasing Dept. to secure samples and further data. After a few weeks with no reply, I again asked for material and data with a similar result. I then asked the Purchasing Agent for the privilege of writing direct, which at that time was against rules, but I was given permission to do so. I got a most cordial reply by return mail from Dr. Baekeland, inviting me to go to Yonkers* where he had

* See chapter 2.

his home and a little private laboratory, telling me he would give me all
information he could and help me in
any application experiments we cared
to make. On arrival, he explained that
he considered his material entirely experimental and that he replied to no
purchasing agents but was anxious to
work with engineers, so that applications could be made in a mutual co-
operative way. I could think of many
applications, such as substituting
Bakelite for shellac in Micarta. A
series of test and applications were begun with almost 100% failures. It
would make quite a catalog to recount
the applications which did not yield
results equal to our going processes using shellac and varnish. Dr. Baekeland
and I worked hard to run down the
causes of failure and to improve the
material and the product. We found
mechanical strength greatly improved
over shellac but dielectric failures and
losses were high. We had trouble in
reaching the desired step in the coating of paper so it would work well in
our machines. It was not long until

one application after another was satisfactory. New applications, not possible with shellac, gave better structures and in a relatively short time we became a large user of Bakelite.

Years later, Dr. Baekeland told Dr. Skinner: "Before I started work on the phenolic resin enigma, I lived a life of a philosopher up the Hudson—messing around in my laboratory. After starting on Bakelite, I never again had any leisure."

Westinghouse requirements outgrew the capacity of the laboratory still, and Lawrence Byck tells the following story about the initial run in the new 500-gallon still at Perth Amboy, making laminating varnish #1 for Westinghouse:

It was a historic event. The first chemist to assist Mr. Rossi had been hired some time earlier. He was J. J. Frank. While the plant was being installed, he was given a course of instructions at the Yonkers Laboratory, where Thurlow, Gothelf, Byck and Taylor crammed into him in a few weeks, all the information they could regarding Bakelite. This was probably all they knew plus a little more. Then J.J. went to Amboy where he did some development work until all was ready. Finally the big day arrived, the first large batch was to be produced. Dr. Baekeland was on hand, Lou Rossi naturally, the first few factory workers, the master mechanic and his helpers, and an assortment of contractor's workmen, millwrights, electricians, etc., who were completing the building and equipment installations. So there was quite an audience. The day ran on into evening. The electric lighting system was not yet complete, so light was furnished by kerosene lanterns hung here and there. Everything went wonderfully well—the resin was reacted, dehydrated, finished and alcohol was added. Dr. Baekeland asked whether the resin was all dissolved or was there possibly a lump of "B" around the agitator. Quick as a flash J.J. grabbed a lantern, swung open the manhole door and held the lantern to it for a look-see. There was a mild Boom, the hot alcohol vapor

Figure 9–1   Dan J. O'Connor, founder of Formica Insulation Company.

ignited; flames spurted from the man-hole. Somebody yelled, "Run every-body," and everybody left by the near-est exit, whether door or window. That is, everybody left except Lou Rossi and the poor chemist. Rossi calmly slammed shut the still's man-hole, then helped J.J. from the scene. Fortunately J.J. was not badly hurt; he lost his hair, eyebrows, eyelashes, mustache, and his job. The batch of varnish was saved.

Dan J. O'Connor (fig. 9–1), a young Westinghouse engineer, was making phenolic rolled tubing by the old shellac process in 1912. Paper was rolled up under pressure on a mandrel (fig. 9–2) while Bake-lite varnish was applied as a binder. This tubing was then baked to harden the varnish. Dan O'Connor conceived the idea of slitting the re-sulting tube and pressing it out flat under heat and pressure for the cure; he thus made the first phe-nolic laminated sheet. A patent was issued in 1918 to Westinghouse for

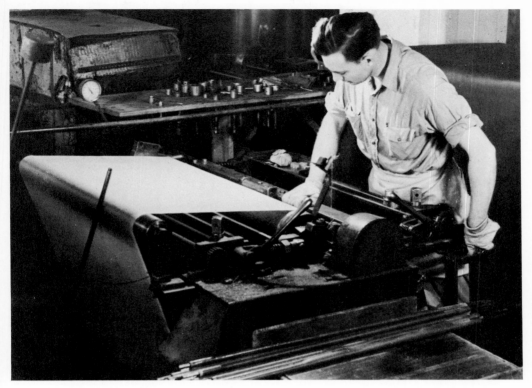

Figure 9–2 An early machine for making rolled phenolic tubing. A sheet of impregnated material travels over a heated roll and, under pressure, is wound on a heated mandrel. Such tubes were first made with a shellac binder. (Courtesy, Formica Insulation Company.)

O'Connor's work, and from this small start developed the very large Micarta operations at Westinghouse and the Formica Company.

Herb Faber, Manager of Insulating Materials Sales for Westinghouse, and Dan O'Connor decided to go into the insulation manufacturing business themselves, and the Formica Insulation Company was formed in 1913 to produce commutator V-rings using laminated insulation to replace the more costly and less desirable mica. The two partners started with homemade screw presses to fill their first order for commutator rings from Chalmers Motor Company.

Because of Bakelite's policy at that time of making their customers licensees, Westinghouse was permitted to make sheets, and Formica was limited to tubes. Dr. L. V. Redman (fig. 9–3) of the Karpen Brothers Company (see chapter 3) agreed to furnish resin that enabled

*Figure 9–3   The first Formica sales conference was held in 1922. In the
top row, third from the right, is Dr. Lawrence V. Redman. At his left
is Dan O'Connor.*

Formica to build a sheet business.
Dr. Redman was most helpful to
Dan O'Connor in solving the early
problems with the laminates.

Frank Conrad, another Westinghouse engineer, developed in 1912
the use of laminated cloth phenolic
for gears (fig. 9–4). They developed a licensing system in 1917 to
collect on the Conrad patents and
in the process formed the American
Gear Manufacturers Association.
General Electric Company's 1914

Miller Gear patents were based on
the Fabroil gear in which cotton
was held in compression between
steel shrouds by steel studs. This
patent was deemed to concern the
laminated phenolic gears since they
were also composed of "textile
fibres, held in compression." This
gave General Electric a licensing
position in the very large automobile timing gear (fig. 9–5) business
and was the start of General Electric commercial plastics sales opera-

Figure 9–4   Frank Conrad invented the use of laminated phenolic materials for gears, and a very large market developed therefrom.

Figure 9–5   The business in automobile timing gear contributed tremendously to the build-up of the laminated plastics in the 1920s and the 1930s. It started when a used-car dealer found that he could quiet a very noisy engine with a canvas Bakelite gear.

Figure 9–6 Laminated rods were first produced by compression molding of rolled-up resin-treated stock.

tions. During the 1920s and 1930s, this was a multimillion dollar business with General Electric Company, Westinghouse Electric Company, Taylor Fibre Company, Formica Insulation Company, Continental Diamond Fibre Company, and Perfection Gear Company carrying a large portion of the load.

The early manufacturing processes followed the shellac method for rolled tubing. Rods were made by rolling up the laminate and then curing in a compression mold as

shown in figure 9–6. Multiopening presses with polished press pans between the laminates were used in the sheetmaking process as shown in figure 9–7. The first sheets were quite small but, as the demand increased, the sheet sizes increased for more economical cutting. Molded laminates were produced in compression molds with hand or machine lay-up of the compounds: Timing gears at one time were molded from machine-punched sectors for the rim with a cotton core

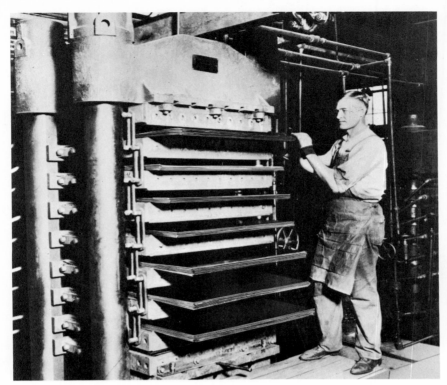

Figure 9–7 Early press for laminated sheet. (Courtesy, Formica Insulation Company.)

for the web plus a molded-in steel mounting bushing. These several components were loaded automatically in hand molds that were machine loaded into the press for the cure.

The automobile breaker (fig. 9–8) has been fabricated for several decades from the laminated phenolics in spite of the many sophisticated molded plastics developed in the interim.

World War I was most instrumental in developing the laminated industry. Dr. Lee de Forest set up specifications for the "wireless" components that upgraded quality control procedures. Westinghouse developed a molded laminated propellor (fig. 5–11) that did not shatter when struck with bullets.

The Signal Corps and the Navy became large users of laminated phenolic sheets for their telephone and wireless components. Curtis Wright, Glen Martin, and other aircraft makers began to use the laminates for pulleys, electrical in-

Figure 9–8 Fabricated laminated insulation parts continue to be used for the ignition breaker switch. This was one of the first applications in the automobile. (Courtesy, Synthane Taylor, Oaks, Pennsylvania.)

sulation, and structural parts. Post-war wireless amateurs in the years 1918 to 1922 developed many new uses for the laminates. These products grew into very high-volume use as radio broadcasts expanded and people were constructing their own receivers completely or by the assembly of factory built components. All of these products produced a great demand for the black, highly polished laminated sheets and laminated tubing.

In 1922, Formica entered into an agreement to put Western Electric Company into the making of their own laminated sheets—and concurrently selling them a considerable volume of laminates. Other high-volume users of the laminates also started to make their own.

The textile industry was another burgeoning market for the laminates. They needed chemical- and abuse-resistance for bobbins, rayon spinning buckets (fig. 9–9), doctor

Figure 9–9 Molded phenolic-filled canvas laminated stock was used for molding these rayon spinning buckets. (Courtesy, General Electric Company.)

blades, picker blocks, etc.—all of which were produced by the laminators.

The coming of the decorative laminates was the next large step upward for the laminators. This was pioneered by Formica. Starting with printed mahogany or walnut-grained panels for the radio industry and marble designs for soda fountains, they moved along to the ·light-colored materials that were a boon to the furniture trade. Tre-

mendous sales volume resulted for dinette table tops and for hotel and restaurant furniture where cigarette burns, household chemicals, and alcohol damaged the natural wood products.

The offering of the thiourea materials in 1931 and the underlayer of metal foil to minimize cigarette damage followed by the urea and melamine colored materials in 1938 put the decorative laminates into the kitchen and in many structural

Figure 9–10 Laminated phenolic strips were selected in 1927 to be used for thermal insulation between the inner steel box lining and the outer cabinet. This switch from wood to plastics by General Electric Company was followed by all refrigerator makers and created a very large market. (Courtesy, General Electric Company.)

places. The 1935 *Queen Mary* used a large volume of decorative laminates—further popularizing these products. Realwood laminated, using resin-bonded wood-veneer surfaces, made possible the use of laminates in fine furniture, offices, etc.

Jack Cochrane of Formica received the 1948 Hyatt Award for his participation in the development of the melamine laminates.

The laminates continued to grow during the depression years with radio parts, the decorative materials, and the refrigerator breaker strip carrying a large share of the load. The breaker strip (fig. 9–10) was pioneered by General Electric Company in its 1927 monitor-top refrigerator to separate the inner and the outer steel shells with a material of low thermal conductivity that was wear resistant. Prior to then, wood had been used, with uncertain life and poor results. During

Figure 9–11 Phenolic-resin-bonded belt duck roll-neck bearing for the steel mills will withstand very high shock loads.

the 1930s, it became economical to mold the entire inner-door integral with the breaker strips, the whole having a white urea surface sheet. These were the big build-up years for the household refrigerator, and many tons of molded laminated material were used. Ultimately after World War II this business was lost to the thermoplastic extruded and thermoformed materials. Roll-neck bearings (fig. 9–11) for the steel mills provided another large and important market taken over by the laminates in the early 1930s, replacing babbit and bronze.

World War II brought a complete sell-out of the existing laminated facilities. The glass-melamine materials solved the circuit-breaker problem and offered very high-strength products to aircraft makers. Wood veneers impregnated with plastics resins (fig. 9–12) were used for the sturdy aircraft propellors and highly stressed parts. The

Figure 9–12 Veneers for aircraft propellers and structural components were vacuum impregnated with phenolic resin.

Figure 9–13 This mechanical fixture performs multiple operations in sequence on the laminated tubes.

Figure 9–14   Instrument panels in World War I extensively used laminated nameplate materials. (Courtesy, General Electric Company.)

P-51 had 88 parts of Formica laminated. Bomb-burster tubes (fig. 9–13) and bazooka barrels were made from laminated tubing. Radio, radar, ships, and all the vehicles employed the laminates. Instrument panels (fig. 9–14) used laminated nameplate materials very extensively. Ball-bearing retainer rings (fig. 9–15) went to laminated tubing to gain phenomenal wear and vibration achievement. Stern tube and rudder bearings (fig. 9–16)

went from lignum vitae to laminated phenolics, gaining extended life.

One of the first high-volume World War II laminated products was the trim tab that was molded by Taylor Fibre Company (shown in fig. 9–17). Coil forms for the many electronics applications were fabricated from the low-loss phenolic laminated tubes as shown in figure 9–18.

The development of postforming

Figure 9–15   Ball-bearing retainer rings fabricated from the high-pressure laminates minimize wear, noise, lubrication, inertia, and malservice.

Figure 9–16   Laminated bearings were used to replace lignum vitae for stern tube and rudder bearings in World War II. (Courtesy, General Electric Company.)

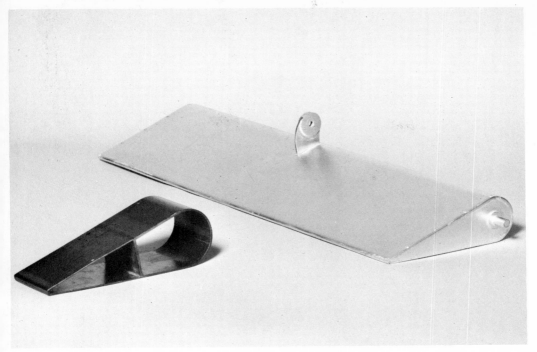

Figure 9–17 One of the first high-volume laminated products of World War II was the trim tab molded by the high-pressure laminating process using phenolic-bonded linen fabric. (Courtesy, Synthane Taylor, Oaks, Pennsylvania.)

Figure 9–18 A radio-coil form is being cut in this World War II production picture.

methods for the laminated materials in 1942 added greatly to their utility. The F-86 plane had 75 postformed laminated components. William Iler Beach, Chief of Plastics Engineers, North American Aviation, Incorporated was given the 1945 Hyatt Award for his pioneering work with postformed laminates.

The largest new market for the high-pressure laminates resulting from World War II was the printed circuit.* The printed circuit came to general attention in 1942 when Dr. A. Ellet and Harry Diamond in the Ordnance Division of the National Bureau of Standards reported the exploration of two-dimensional printing of electronic parts for the Army's V. T. Fuze.

The first practical work on the etched copper laminated circuitry was done at Radio Corporation of

* S. F. Danko, "Printed Circuits and Microelectronics," *J. Am. Inst. Elec. Engrs.*, May 1962.

America (RCA) in 1947. Mr. D. Mackey was responsible for the electrical and mechanical design, establishing guidelines for materials and process development. Developmental work was undertaken by the chemical and physical laboratories under Mr. C. Eddison, and Dr. Otis D. Black was the staff member assigned to the project. Their initial project was a TV tuner using "printed" inductors for general use by television manufacturers. Halli-crafters Company pioneered this tuner and publicized it widely. The initial work in developing suitable copper-clad laminated phenolic plastics was done by Synthane Corporation of Oaks, Pennsylvania. The development involved working with American Brass Company to gain a suitable copper foil and with Industrial Tape Company to get a workable adhesive. This has led to complete revolution in electronics product manufacturing—to

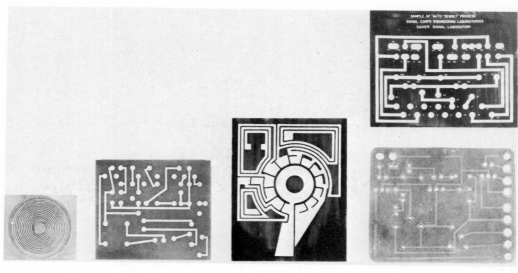

Figure 9–19 These early experimental printed circuits illustrate the various techniques of circuit fabrication showing (left to right) stamped circuit, etched foil laminated, pressed powder circuits, etched circuit sandwiched between phenolic sheets, and (top) inlaid circuits. The first etched circuit was conceived by Dr. O. D. Black and Don Mackey of Radio Corporation of America in 1947. Synthane Corporation made their initial circuits. (Photo, courtesy of Signal Corps Engineering Laboratories, 1951.)

the great benefits of the plastics industry, as depicted in figures 9–19 and 9–20.

The Signal Corps established a miniaturized group at the U.S. Army Signal Research and Development Laboratory in 1946, which resulted in their Auto-Sembly process and a one-shot solder-dip system developed by Danko and Abrahamson. Concurrently, J. Beck was developing copper forming tech-

niques, Henderson-Spalding in London, England, were producing etched circuits on flexible plastics based on Paul Eihler's U.S. patents, and Franklin Airloop Corporation was embossing copper coils on plastics. R. L. Swiggett formed Photocircuits Corporation for the production of etched circuits in 1951. The first printed-circuit radio was introduced in 1952.

From this small start, the etched

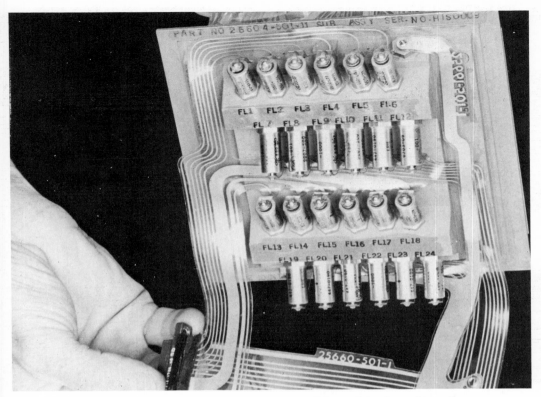

Figure 9–20 Extreme environmental conditions often have been the sources of markets for fluorocarbons. This flexible printed circuit on Teflon FEP film carried up to 18 amperes in some circuits and is wrapped around one component of the Minuteman III computer.

laminates and plastics film circuity took a major position in electronic circuitry.

The phenolic-resin, bonded-plywood development followed World War I. At that time, blood albumen and casein glues had been used and found to be unreliable in aircraft. Gordon Brown of Bakelite Corporation worked with Forest Products Laboratories and developed water-soluble resins that gave excellent high-strength bonds. The plywood industry then changed at a fast rate to synthetic resins with phenomenal product and market gains. These developments made possible the large use of wood in the gliders and training planes of World War II.

In the early stages of World War II, plywood was molded into three-dimensional shapes using phenolic and urea resins. Several new processes such as the Duramold, Vidal, Timm, and Haskelite methods were devised that made use of rubber bags to apply the required high pressures. The coming of the low-pressure reinforced plastics made obsolete many of the initial wartime developments.

Reinforced Plastics

Prior to the development of the polyester resins and their laminated products, high molding pressures had been used for all of the laminates. The new laminates molded with the polyesters could be processed with little or no pressure. They were called low-pressure laminates, and the older materials called high-pressure laminates.

For two decades the name reinforced plastics was accepted to designate the structural low-pressure laminates and the filament-wound products. Some confusion has resulted from the 1967 name change of the SPI Reinforced Plastics Division to the RP Composites Division. Shellac and all of the thermosets are composites that were used from the start with filler reinforcement; many fillers serve in the thermoplastics molding compounds for added strength and other desirable properties.

Paper boats, bonded with adhesives are the first recorded reinforced plastics products. Mr. George A. Waters of Troy, New York, built a laminated paper boat in June 1867 that is a forerunner of the present-day reinforced plastics boats. He took a wooden shell, 13 in. wide and 30 ft. long, as a mold and covered the entire surface of its bottom and sides with small sheets of manila paper, glued together and superimposed on each other so that the joints of one layer were covered by the middle of the sheet immediately above, until a shell of paper $\frac{1}{16}$ in. in thickness was built up. This shell was then fitted with a wooden frame consisting of a lower kelson, two inwhales, the bulkheads; in short, all of the usual parts of a wooden shell except the timbers or ribs. All surfaces were then waterproofed with contemporary varnishes. Subsequent boats made by this process used continuous sheets of paper from stem to stern. These boats were very popular and successful. Noteworthy is the fact that four were purchased by the U.S. Naval Academy in 1868.

The development of glass filaments started in 1930 when a molten glass rod was being used to apply lettering on a glass milk bottle. Further studies led to patents in 1936 by G. Slayter and J. H. Thomas. Owens Corning Fiberglas Company was formed in 1935 to make and sell these fibre glass products. Glass wool for insulation and filters was the first product with synthetic resins being used as a packaging material.

Carleton Ellis did his original work on the polyester resins in 1933; the Ellis-Foster 1937 patents covered a production method for making them. The first glass-fibre-reinforced polyester products were made in 1938. Polyester resins CR-38 and CR-39 were introduced commercially in 1940 by Pittsburgh Plate Glass Company. By 1942 American Cyanamid Company; E. I. du Pont de Nemours and Company, Incorporated; Libby Owens Ford Company; Plaskon Company; and Marco Chemicals were all producing polyester resins. Flame-retardant polyester resins were developed by P. Robitscheck and C. Thomas Bean, Jr., of Hooker Chemical Company.

The Air Force issued a contract in 1942 to Marco Chemical Company of Linden, New Jersey, where

Irving Muskat continued the development of low-pressure resins which he had studied at Pittsburgh Plate Glass Company. Another research contract was issued to Owens Corning Fiberglas Company to expand the art and capability of these new high-strength plastics, which could be molded with little or no pressure. The Air Force put great pressure behind this program to get an alternative for magnesium, plywood, and aluminum. By the end of 1942, important low-pressure laminated and molded parts were being produced by Uniroyal, Goodyear, Formica, Boeing, Douglas, Grumman, Westinghouse, General Motors, General Electric, and Swedlow. Production became tremendous overnight.

The first fibre glass boat of record was molded in 1942, and the Navy secured its first boat in 1946. Laminated sheets of glass fibre and CR-38 and CR-39 resins were soon

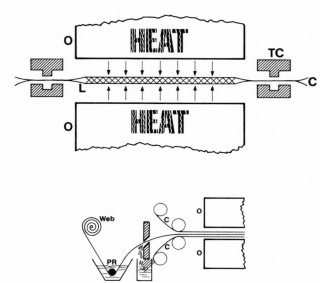

Figure 9–21 First continuous low-pressure polyester laminates. The polyester resin laminate is carried on two cellophane sheets, one top, one bottom. The cellophane sheet becomes taut when heated. Drawing B shows the process. L = polyester resin laminate; O = electrically heated oven; TC = tender frame clamp; C = cellophane carrier sheet(s); PR = polyester resin dip tank.

offered by Pittsburgh Plate Glass Company for aircraft, boat, and automobile parts; CR-39-bonded laminates by Goodyear Aerospace Corporation were used for aircraft fuel-cell backing.

Continental Can Company in 1942 was the first to set up a continuous low-pressure polyester-laminating process. Three-ply polyester-bonded fibre glass cloth (fig. 9–21) was laminated for military aircraft by combining the treated glass fabric and curing it continuously between sheets of Cellophane.

Studies at Wright Air Development Center in 1943 approved structural sections with high-strength polyester-glass-laminate faces and a low-density core material which led to many high-strength aircraft and missile components. The BT-15 airplane with fuselage of this sandwich construction was first flown in March 1944, with great success.

Figure 9–22   The filament-winding machine is used to produce very high strength tanks and products from the reinforced plastics.

Introduction of the epoxy resins in 1946 (chap. 5) added strength to the reinforced plastics. The filament-winding process (fig. 9–22) was also developed in 1946. R. E. Young at M. W. Kellog built the first filament-winding machine and produced rocket motor cases and filament-wound pipe (fig. 9–23). The Glastic Corporation was founded in 1946 to concentrate on the glass reinforced molded, and laminated plastics.

A rigid fibre glass shelter was started in 1946 to protect the observers who were tracking balloon-borne transmitters. These shelters, now called radomes, had an inside diameter of 14 ft 8 in. to provide rotational space for the SCR-658 radiosonde direction-finding apparatus. H. W. Rahmlow was the engineer responsible for this pioneer radome. The initial structure was installed for the Weather Bureau at Sault Sainte Marie, Michi-

Figure 9–23   These fiber-glass-reinforced launch tubes for the Redeye Missile were made by the filament-winding process by Lamtex Industries in the 1960s. (Courtesy, Koppers Company, Incorporated.)

gan, in 1949 and is shown in figure 9–24. Similar radomes have been installed all over the world with no failures, while other structures were often demolished.

A 12,000-lb reinforced plastics fairwater was installed on the submarine *Halfback* in 1953 and the Boeing B-57 introduced the use of phenolic fibre glass ducting.

The Corvette body of 1955 produced by Molded Fiber Glass Body Company started the use of polyes-ter glass reinforced plastics for automobile bodies (figs. 9–25, 9–26, 9–27). This giant step matured into 300 million lb of reinforced plastics in automobile components in 1970. The 1970 applications included front-end valances, front-bumper closures, headlight housings, parking and signal lamp housings, outer hoods, truck hoods, cold-air intakes, floorboards, deflectors, spoilers, fender extensions, turn-lamp housings, instrument panels, dash-

Figure 9–24   *This first radome was started in 1946 for tracking balloon-borne transmitters. It was made of polyester and fiber glass in white-metal molds and installed at Sault Sainte Marie, Michigan. (Courtesy, N. W. Rahmlow.)*

Figure 9–25   *The first production automobile body to utilize plastics was the 1955 Corvette, molded of polyester resin and glass fibers. (Courtesy, Molded Fiber Glass Body Company, Ashtabula, Ohio.)*

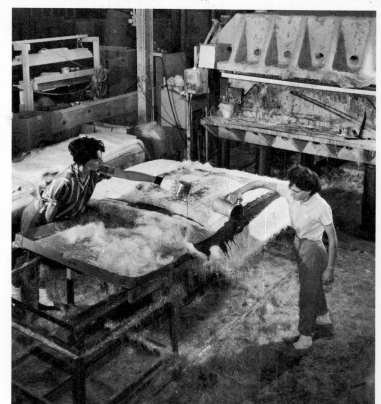

Figure 9–26 Hand lay-up was used in the molding of the 1955 Corvette. The resin is being applied to the glass fibers in this view. (Courtesy, Molded Fiber Glass Body Company, Ashtabula, Ohio.)

Figure 9–27 The workers are removing the flash from a front upper section of the Corvette body that has been molded of reinforced plastics. (Courtesy, Molded Fiber Glass Body Company, Ashtabula, Ohio.)

Figure 9–28   The polyester glass pharmaceutical trays of 1953 started a revolution in industrial trays. (Courtesy, Molded Fiber Glass Tray Company, Ashtabula, Ohio.)

boards, gear-shift consoles, trim panels, arm rests, heater and air conditioner housings, plus numerous small parts. Polyester glass materials took over the industrial tray market in 1953 (fig. 9–28).

Ablation studies in 1955 on phenolic glass and phenolic asbestos pointed the way to solutions of the missile-reentry problem. The first plastics reentry nose cone was built in 1956 and flown successfully on the Vanguard missile.

Studies at Texaco Incorporated in 1959 indicated the large potential from high-strength boron fibres in reinforced plastics. In 1960, Piper Aircraft Company flew a reinforced plastics airplane with polyester glass skins on a paper-honeycomb core. Concurrently Dow Chemical Company started building the Windecker glass-epoxy-reinforced plastics wing for low-wing planes. This was constructed as a single structure with a rigid

Figure 9–29   This new helicopter driveshaft marks the first use of boron-epoxy composite materials. A great saving in weight was achieved by this "next-generation" helicopter driveshaft. (Courtesy, Whittaker Corporation, Los Angeles, California.)

urethane foam core. The development of carbon fibres for the reinforced plastics in the late 1960s is expected to become rated as one of the great technical developments of the 1960s. One serious study reports that by 1980 70% of all manufactured products will be made with carbon-fibre reinforced plastics. The initial developmental work on this promising new material was done by the British Royal Aircraft Establishment and Union Carbide

Corporation. Use of boron-fibre reinforced plastics found important applications, (fig. 9–29) in the late 1960s.

All of these pioneering developments gave birth to many commercial products such as the reinforced plastics bathroom exhibited at Canada's Expo 67 and as marketed in the United States by Crane Company and others (fig. 9–30). Fishing rods, pipe, tennis racquets, fur-

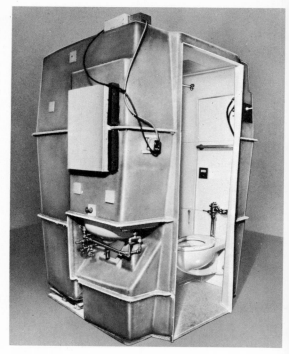

Figure 9–30   This first fully equipped packaged plastics bathroom was introduced by Crane Company in 1967. It was designed as a modular unit that can be shipped in a compact package and assembled on the site by standard practices. (Courtesy, Crane Company, New York, New York.)

Figure 9–31   Filon developed this continuous-sheetmaking machine for transparent laminated polyester glass sheets in 1968. The unit will produce corrugated and textured sheets. (Courtesy, Vistron Corporation, Hawthorne, California.)

Figure 9–32 Translucent polyester panels are shown here being installed in the new building at 666 Park Avenue, New York City. Precision factory-made panels are 5 ft wide and 14 ft long and weight 1.5 lb per sq ft. (Courtesy, Kalwall Corporation, Manchester, Connecticut.)

niture, trucks, railway cars, tanks, and a wide variety of structural and insulating components of reinforced plastics moved into almost every market previously served by alternative materials. The building industry began its big switch to plastics in the 1960s (figs. 9–31 and 9–32), along with many other structural and decorative product applications that will mature in the upcoming plastics age.

10

Plastics Industry Services

Many of the established organizations which contributed to the growth of the plastics industry during the formative years have now been replaced by internal organizations. Particularly helpful in the early years were the following organizations: National Electrical Manufacturers Association, American Society of Mechanical Engineers, Institute of Radio Engineers, Plastics Material Manufacturers Association, Synthetic Resin Manufacturers Association, American Society for Testing and Materials, American Chemical Society, Society of Automotive Engineers, Technical Association of the Pulp and Paper Industry, and many local technical groups. The need for internal organizations brought together many of the early local industry groups into the Society of the Plastics Industry, Incorporated, and the Society of Plastics Engineers, Incorporated.

The Society of the Plastics Industry, Incorporated (SPI)

SPI was formally organized on May 11, 1937, when it was incorporated in the State of New York. There were 130 molders, materials suppliers, and representatives of the plastics industry who attended the annual get-together at Shawnee-on-the-Delaware, May 24 and 25, 1937. At this meeting, the Society of the Plastics Industry, Incorporated, started its work. The primary purpose as enunciated at the time was to build up good fellowship in the industry. The originators recognized that there was much to be gained by unselfish cooperation on noncontroversial matters such as new uses for plastics, new types of materials, and developments of the future that were expected to affect the industry. The stated purpose of SPI is to provide leadership for the responsible advancement of the entire plastics industry. The Society, as a not-for-profit organization, in furtherance of this basic purpose, has the following objectives:

to represent and serve as the official spokesman for the plastics industry in the United States;

to promote the effective use and application of plastics, consistent with the public interest;

to provide and stimulate authoritative organized research, education, and information within the industry and with other industries, government bodies, and interested organizations;

to provide means for mutual

communication and organization of groups within the industry to initiate and pursue programs of common technical, marketing, or management interest;

to mobilize and finance voluntary staff and professional expertise to provide the required range of services to members; and

to maintain liaison and cooperate with other plastics and allied trade and professional associations in the United States and in other countries throughout the world.

The Board of Directors and Officers of the newly formed Society in 1937 were: Prescott Huidekoper, President; Ronald Kinnear, Vice-President; William Kelly, Secretary and Treasurer; G. Victor Sammet, Alan Fritzsche, Gordon Brown, and James L. Rogers. Initially the Soci-

ety's work force was the many volunteers who devoted endless hours and sums of money to building SPI. Eventually its activities and interests became sufficiently broad so that the services of a full-time secretary were necessary. The first Secretary, hired in the spring of 1941, died unexpectedly within four months. Wm. T. Cruse (fig. 10–1) who was a member of the Board of Directors, succeeded to the office in October of 1941 and served until 1967. He was succeeded as Executive Vice-President by Ralph L. Harding, Jr., who now directs all the Society's many activities.

SPI's total staff in 1970 numbered 50, including Toronto, Chicago, and Los Angeles offices. The Society cooperates with other trade societies and associations to avoid duplication of efforts and to achieve maximum benefits in connection

with problems of the industry.

In 1937 there were only 12 basic types of plastics compared to 20 today. SPI had 75 charter members in 1937 compared with today's 1,335 member companies in the United States, Canada, and 29 other countries, and 1,351 individual members. Now, there are also offices in Los Angeles and Chicago and a Canadian section known as the Society of the Plastics Industry (Canada) Incorporated that has 137 member companies and 65 individual members. SPI has also 100 working committees, divisions, or other groups which are concerned with every facet of the plastics industry and which sponsor or identify with 350 different meetings, symposiums, or conferences annually.

Society of Plastics Engineers, Incorporated (SPE)

In the early days of the plastics industry, many local groups such as the Plastics Engineers Association, the Plastics Club of the United States, the Pacific Plastics Society, etc., served the need for an exchange of ideas and the hunger to be better informed about plastics materials, machinery, and methods.

SPE started very modestly in 1941 at a golf dinner where a group of sales engineers gathered as had been their custom in previous years for fellowship and an exchange of views. A decision was reached to formalize the organization and to plan regular programs, speakers, and meetings. The name Society of Plastics Sales Engineers, Incorporated, was initially adopted, and articles of incorporation were filed for the purpose of cooperating, aiding, and effecting any commercial or industrial betterment pertaining to designing, styling, standardizing, and promotion of the use of plastics through distribution of descriptive matter or personal contact or in convention in any manner that may be educational. The name was soon changed to Society of Plastics Engineers, Incorporated, and provisions were made for forming chapters in other locations.

One year later, the membership had expanded to 120 persons. Today, in 1970, there are 16,500 members of SPE and 67 sections. The first Society publication, *The Commentator,* edited by Fred Conley, was issued in May 1942. It was a mimeographed bulletin of 24 pages with a colored masthead.

The first Regional Technical Conference was held in Chicago in 1943, and 6 papers were presented on Dielectric Preheating to more than 300 persons. The first Annual Technical Conference was the symposium and exhibition held at the Horace H. Rackham Memorial Building in Detroit on October 8, 1943. It was a tremendous success which attracted 59 exhibitors and 1,775 registered visitors. No national meetings were held in 1944 or 1945 because of wartime restrictions on travel.

The SPE statement of purpose reads:

The objective of the Society shall be

to promote the scientific and engineering knowledge relating to plastics. The objectives shall be achieved by:

Technical meetings at which formal and informal seminars are conducted for the presentation of scientific and engineering reports and papers and for the discussion of problems related to the objective of the Society.

Scientific and engineering reports and papers and an Official Publication devoted to the dissemination of technical information relating to plastics.

Encouraging engineers and scientists to exchange technical information relating to plastics.

Cooperating with and encouraging educational institutions to establish technical and scientific sources relating to plastics and to maintain high standards of technical education.

The work of the Technical Volumes Committee has resulted in useful contributions to the plastics literature and technology. Many of these works could not have been completed by one individual.

In the educational field, SPE has sponsored educational programs in cities, schools, and colleges throughout the country. The annual Technical Conferences have provided the outstanding yearly achievement of SPE. At these conferences, a selection of useful, contemporary, and advanced technological and scientific data is presented. The vast expansion of the science and technology of plastics is making it increasingly difficult for the individual to keep abreast of the literature and practice—even in specialized segments of the industry.

The Society of Plastics Engineers satisfied the great need for the expansion of the technical horizons, for the intercommunication of various specializations, and the development of standard procedures and quality. SPE has provided also the all-important professional recognition and prestige for the plastics engineers.

Plastics Institute of America (PIA)

In the early growth years, there were no educational programs having value for the processors of plastics. The first coordinated effort that went beyond the courses in colloidal chemistry and formation of polymers that were taught in a few schools was the establishment of the Polymer Institute at Brooklyn Polytechnic Institute under Dr. Herman Mark in 1945. Concurrently and independently, the Plastics Program was started at Princeton University under Professor Louis F. Rahm. The Polymer Institute concentrated exclusively on pure science, and the Plastics Program was directed to the engineering aspects of the polymers.

The Plastics Program contributed greatly to the processing of plastics know-how and caught the attention of the Society of Plastics Engineers—leading up to the formation of its National Education Committee under Professor Louis F. Rahm. A similar committee and program was formed by SPI. These groups made many studies and publicized the need for educational planning by the plastics industry.

Professor Rahm recommended a feasibility study by SPE in 1956 leading to an industry-sponsored research and development program. Concurrently Dr. Gordon Kline published a similar plea in *Modern Plastics*. A formal recommendation covering the development of the Plastics Institute was recommended and accepted by SPE Council in 1957. The SPE committee found little support from industry and strong opposition from SPI and MCA and then decided to go ahead alone with a plan based on the very successful Textile Research Institute at Princeton. Jerome Formo and Professor Louis F. Rahm carried the interim torch for the Plastics Institute with many surveys and studies in industry and the appropriate educational institutions.

Mr. Jules Lindau, III, took over the committee chairmanship and secured a charter on June 6, 1961, for a scientific, charitable, and educational institution to be known as the Plastics Institute of America. Jules Lindau, III, became the first Chairman of the Board of Trustees, and Thomas Zawadski was engaged as Executive Secretary, starting the formative work on October 15, 1961. An agreement was reached in March 1962 with Stevens Institute to house the PIA and a plan of study was developed at the BS, MS and PhD levels. With loans from SPE, the Executive Secretary, and the Chairman, PIA carried on and developed an income equal to expenses by March 1963.

In subsequent years the hard work of SPE, Professor Rahm; Jerry Formo; Jules Lindau, III; Thomas Zawadski; and the many committeemen has paid off. All opposition has been withdrawn, and industry came in with much support and help. PIA today speaks for itself by its service to industry and the training of plastics engineers and scientists. Dr. A. W. Meyer was appointed Executive Secretary in 1967, bringing in industrial experience and strong leadership.

The Plastics Education Foundation

Recognizing the need for expanded training programs at all work levels, The Society of Plastics Engineers Incorporated and the Society of Plastics Industry Incorporated formed an industry program in 1970 for the development of manpower needs. The central purpose of the Plastics Education Foundation is to achieve improved quality and broadened availability of educational resources, facilities, curriculums, and teachers in plastics science, engineering, and technology for people of all ages from high school through their adult years, to offer them challenging career opportunities and provide the plastics industry with continuing, qualified talent at all levels of capability as needed to maintain the industry's dynamic growth, bolstering but not duplicating efforts of those organizations and agencies which already function effectively in any segment of this broad field, while stimulating new services wherever they may be needed. Early plans include career guidance films, guidance books, and study programs for trade and vocational schools, junior and community colleges, and the universities.

The Plastics Pioneers Association

Mr. Herbert Spencer, having heard other SPI members complaining that the industry meetings were growing so large and had so many newcomers that it was increasingly impossible to find their friends, suggested starting an oldtimers branch of SPI. Several informal meetings were held, resulting in the development of a formal organization in 1942. The initial officers were Herbert Spencer, President; Nick Backsheider, Vice-President; and Hans Wanders, Treasurer. Hans Wanders continued to serve as Secretary-Treasurer until 1970. The first Board of Governors included Jim Neal, Al Manovil, Ed Bachner, Alan Cole, Garson Meyer, and Doug Woodruff.

The initial membership requirement stipulated fifteen years of employment in the plastics industry and membership in SPI. The length of service has been extended as time passed. In 1953, a series of tape recordings were made by many pioneers to record the early days of plastics and some of the unofficial happenings. These tapes are on file at the Smithsonian Institution for

the use of future historians. Much basic data and color for this book were taken from these tapes.

The Plastics Pioneers Association is strictly a social group with annual meetings (fig. 10–2) and contributions to the scientific and educational needs of the industry.

The Trade Magazines

MODERN PLASTICS   Prior to 1926 no plastics magazines were published in the United States. *Rubber World* and the chemical magazines published bits of news about these materials. One day in 1926, Sylvan Hoffman, a publisher, and Irving Goldberg, a plastics fabricator, were riding together on the New Haven Railroad when Goldberg remarked that there was no trade paper published for his business. This comment caused Hoffman to start an investigation with Alan S. Cole, one of his salesmen.

Alan Brown of Bakelite Corporation gave them encouragement, and the Charles Burroughs Company signed on as the first advertiser. Alan Cole then toured the country, visiting principal machinery and materials makers, gaining 25 advertisers. Bakelite Corporation bought the back cover and General Plastics Company (Durez) bought the second cover.

The original magazine *Plastics* was first published by Neckwear Publishing Company in October 1925 with Sylvan Hoffman as Publisher, Alan S. Cole, Business Manager, and Karl Marx, Editor. William Gruen replaced Marx at a later date. The name was changed from *Plastics* to *Plastics and Molded Products* to gain the support and interest of the processors.

In 1930 Mr. Harris Dobble, a trade paper broker, brought in Mr. Robert C. Gilmore, Sr., who wanted to buy the magazine for his son who was graduating from school in journalism. It was sold for $60,-000 and Gilmore took over in March 1933. The depression was just coming in, and Gilmore became bankrupt and gave the magazine to Williams Haynes, a publisher of a directory called *Chemical Markets,* in return for a $50 per week job. The magazine was then published by Plastics Publications.

In the interim, Alan S. Cole had left Hoffman and joined Breskin and Charlton, who then published *Modern Packaging*. At Cole's urging, Charles Breskin contacted Williams Haynes and purchased *Plastics and Molded Products* for $9,000. To conform with *Modern Packaging,* the name was changed to *Modern Plastics*. Earl Lougee then became Editor and, from that point forward, *Modern Plastics* expanded rapidly. The first *Modern Plastics Encylopedia* was published in 1936.

Modern Plastics and *Modern Packaging* were sold to the McGraw-Hill Company in 1963, and a new era started. Today *Modern Plastics* reports and interprets new developments in plastics markets, processing, and fabricating technology, application and design concepts, technical and engineering research, machinery and equipment, plastics materials, and chemicals and plastics industry structure.

PLASTICS WORLD The most substantial market for plastics initially was in the electrical field. *Electrical World* was then the leading publication. Once each year, Bakelite Corporation carried a 24 to 32 page insert; they paid for the first and last pages and the center spread. C. William Cleworth was assigned to coordinate the sale of advertising space for the intermediate pages to the fabricators and molders of plastics. These users of Bakelite bought full pages of the insert, which was printed on colored coated stock with a standard border design. Bakelite Corporation purchased 20,000 reprints annually for their sales promotion.

Mr. Cleworth joined the Haywood Publishing Company in Chicago in 1937 as Vice-President and Publisher. They started a glamorous monthly called *Packaging Parade,* which brought Mr. Cleworth into the nonelectrical markets for plastics that had begun to displace metal, paper, and glass for packaging.

With this broad acquaintance with men and businesses in the plastics industry, Mr. Cleworth formed Cleworth Publishing Company and initiated *Plastics World* in 1942.

At that time, the pioneer plastics magazine had a circulation of 6,000, made up largely of people within the industry. Mr. W. P. Woodall, who had been manager of McGraw-Hill's direct-mail department, was engaged to prepare a circulation list for *Plastics World,* and he chose departments that were centered around users or potential users of plastics among leading manufacturers. He obtained the names of persons responsible for design, product engineering, manufacturing, etc. Woodall's circulation list numbered 25,000 persons and brought *Plastics World* to the greatly enlarged potential market for plastics materials and products.

Herbert R. Simonds, who had been on the editorial staff of *Steel Magazine* for which he had written a series of articles on plastics, was engaged as Editor. He was then publishing *Plastics News Letter,* which was acquired along with Breskin's *Plastics Bulletin* at a later date and combined into *Plastics World,* first issued in April 1943. It was a 16-page tabloid with a guar-

anteed controlled circulation of 25,000.

Editorially, *Plastics World* covered developments and applications for plastics materials, machinery, and methods with briefly illustrated items. New literature was reviewed. George W. Rhine joined the staff as Managing Editor in 1945. He soon became Editor and continued as long as publication was done by Cleworth Publishing Company.

In the years after World War II *Plastics World* replaced the tabloid format with extensive and fully illustrated articles covering the business problems, new materials, machinery, methods, and markets.

Charles Cleworth, son of the founder, who had spent several years in various departments of *Plastics World* became its Publisher in 1964. Cahners Publishing Company purchased the magazine in late 1964 and assumed its publication in February 1965.

SPE JOURNAL The Society of Plastics Engineers, Incorporated issued its first publication, *The Commentator*, in 1942. It was edited and published by the first president of SPE, Fred Conley. Publication of a magazine was planned in 1943 but delayed because of wartime paper shortages. An interim news bulletin was circulated by Phil Robb until early 1944.

The *SPE News Bulletin* was published quarterly in 1945 and grew into a printed magazine in 1946 plus a 36-page directory issue. Initially it was authored, edited, and published by L. S. Shaw, Paul Reed, and Dr. Jesse H. Day. Its name was shortened to *SPE News* in 1946, and Dr. Day continued as Editor for the next 13 years. During this period when SPE was struggling to keep alive while maintaining ambitious plans for the future, Mrs. Jesse H. Day served as SPE Executive Secretary, and she contributed greatly to the magazine.

In May 1949, the Council selected the name *SPE Journal* for the title of this publication. When Dr. Day retired as Editor in 1958, the publication office was moved from Athens, Ohio, to the General Office in Stamford, Connecticut. A long series of editors followed leading up to George Smoluk, the 1970 Editor. The *SPE Journal* today is devoted to the reporting of advanced plastics technology, the best data from the annual technical conferences, and the news of the society. It is the official publication of the Society of Plastics Engineers, and its August issue is a directory with a complete roster of members.

PLASTICS TECHNOLOGY Recognition of the growing impact of plastics on the rubber industry and the common market that they served in the early 1950s caused *Rubber World,* a Bill Publications magazine to add a department called "Plastics Technology."

The Editor of *Rubber World*, Robert Seaman, and an associate, Mr. Art Merrill, converted the department in 1955 into a magazine called *Plastics Technology*. They were prompted in this action by the accelerated growth of the processing industries.

In recognition of the trend to captive plant plastics processing in the early 1960s, *Plastics Technology* developed a definitive market study in terms of number of plants and personnel involved in the various types of processing industries served and their geographical location.

With this quantitative audience definition, *Plastics Technology* was oriented to process, production, and manufacturing engineering and became devoted exclusively to serving this plastics engineering function. Market-research data and a definition of trends within the industry

are some of the important goals of this publication.

Plastics Technology is published by Bill Communications.

PLASTICS DESIGN AND PROCESSING *Plastics Design and Processing* was initiated in 1961 by Lake Publishing Corporation with Lincoln R. Samelson, Publisher. This magazine concentrates exclusively on serving the interests of those who use, specify, and buy plastics materials, products, processing equipment and related supplies. None of its circulation and editorial content is directed to marketing men or suppliers to the industry.

Plastics Design and Processing's major goal is to inform and instruct readers so that they may develop, improve, and manufacture plastics products and components effectively, efficiently and economically.

PACIFIC PLASTICS *Pacific Plas-*

tics was the early West Coast publication which was published by Miller Freeman Publications of California. The December 1948 oldtimers number carried a very fine historical summary of the start of plastics in the West. *Pacific Plastics* was then the official publication, Pacific Coast Section SPI.

WESTERN PLASTICS Western Plastics was founded in 1954 by Frederic M. Rea, owner of Western Business Publications. It began publication as a monthly news and technical journal in October 1954 to serve the plastics industry and plastics-using industries in the eleven Western states, Alaska, Hawaii, and Texas. The initial circulation of 4,000 covered Western plastics processors and a selected number of other Western industries. Circulation has since grown considerably and includes a cross-section of other industry in addition to plastics processors.

An annual directory of the Western plastics industry was developed and published as part of the January 1956 issue. It listed a few hundred companies and totaled only 68 pages. By 1962 the directory had grown sufficiently in size and significance to make the directory a separate 13th issue each year. The April 1968 issue was devoted to the plastics pioneers and their businesses; this is an excellent historical reference issue.

On October 29, 1970, *Western Plastics* was sold to Breskin Communications, Incorporated, of Scottsdale, Arizona, headed by Theodore B. Breskin, with Fred Rea remaining as Publisher.

House Organs

Two house organs, *Bakelite Review,* published by Bakelite Corporation, and the *Durez Molder,* published by Durez Plastics and Chemical Company, made most valuable contributions to the growth of plastics in the 1930s and 1940s. During this period of tremendous growth they published reports on new products, materials, and methods that were most helpful to the molders of plastics. During the 1930s, these two publications were the real source of mold design and processing techniques.

Industry Awards

Several awards have developed over the years to recognize and reward outstanding services to the plastics industry.

The John Wesley Hyatt Award was established by Hercules Powder Company in 1941 to cite "the individual whose contribution has best served to advance the usefulness of plastics during the preceding year." The award was discontinued in 1953. The recipients of the Hyatt Award were noted previously in the appropriate position in this volume (see Table of Hyatt Awards).

The SPE International Award in Plastics Science and Engineering was established in 1962 to stimulate

Figure 10–3   The SPE International Award in Plastics Science and Engineering is presented annually to an individual who has made fundamental and outstanding contributions to plastics science or engineering or both.

Hyatt Awards

Recipient	Year	Recipient	Year
Dr. Donald S. Frederick	1942	Frank H. Shaw	1943
Dr. Stuart D. Douglas	1944	William Iler Beach	1945
Virgil E. Meharg	1946	Paul D. Zottu	1946
Dr. John J. Grebe	1947	R. R. Dreisbach	1947
John S. Cochrane, Jr.	1948	Dr. George T. Felbeck	1949
G. M. Powell	1950	James T. Bailey	1951
Palmer W. Griffith	1952	Dr. Howard L. Bender	1953

and encourage fundamental contributions in plastics science and engineering throughout the world; to acknowledge outstanding achievements by honoring distinguished scientists and engineers; to disseminate widely among plastics scientists and engineers the technical information of the Award Lectures. The International Award is shown in figure 10–3.

The Modern Plastics Awards were given from 1936 through 1940 and after World War II in 1946; the award plaque is shown in figure 10–4. These awards were distributed to persons contributing to the development and production of outstanding plastics products of the previous year. The awards were made to the designer, producer, and

materials maker and stimulated a real effort for newer and better uses for the plastics.

The Bachner Award, depicted in figure 10–5, for outstanding achievement in the industrial application of molded plastics was established in the fall of 1957 by Chicago Molded Products Corporation and is administered by the Bachner Award Committee composed of individuals who are independent of the company. The Award was created as a testimonial to the work of all who have

Figure 10–5 The Bachner Award trophy for outstanding achievement in the industrial application of molded plastics also honors Mr. Edward F. Bachner, Sr., one of the greatest of the pioneer molders.

International Awards

Recipient	Year	Recipient	Year
Dr. Herman F. Mark	1962	Dr. Giulio Natta	1963
Dr. Carl S. Marvel	1964	Dr. Turner Alfrey, Jr.	1965
Mr. J. Harry DuBois	1966	Dr. Paul J. Flory	1967
Dr. Raymond F. Boyer	1968	Dr. Richard S. Stein	1969
Dr. Arthur V. Tobolsky	1970	Dr. Albert G. Dietz	1971

assisted in the growth of the industry—to encourage and promote continual growth. It is, further, a testimonial to the five Bachner brothers who formed the company in 1919 and who have contributed significantly to the growth of the plastics industry.

A Bachner Award Competition is held every two and one-half years concurrent with the Society of the Plastics Industry's National Plastic Exposition and Conference.

SPE Educator Award: This award was initiated in 1968 and was first given to Professor Louis F. Rahm for his vision and actions in the creation of the Princeton Program and his work to develop the Plastics Institute of America.

Standards For Plastics

H. C. Gunst
Union Carbide Corporation

Standardization for plastics began with the materials which preceded them. Very early in the history of the plastics industry, various technical committees of the American Society for Testing and Materials concerned with the so-called traditional materials turned their attention to the then new plastics and began work on testing methods and specifications for them. Plastics increased in importance to those industries first concerned so that work on them continues to this day in ASTM committees D-1 on Paint, Varnish, Lacquer, and Related Products; D-2 on Petroleum Products and Lubricants; D-8 on Bituminous and Other Organic Materials for Roofing, Waterproofing, and Related Building or Industrial Uses; D-9 on Electrical Insulating Materials; D-11 on Rubber and Rubber-like Materials; D-16 on Industrial Aromatic Hydrocarbons and Related Materials; and, more recently,

F-2 on Flexible Barrier Materials, among others.

By the midthirties, plastics had gained acceptance in their own right so that, in 1937, ASTM Committee D-20 on Plastics was organized and, during the almost 30 years which have been passed since then, D-20, like the industry which it serves, has grown so that today it is one of the largest and most active technical committees in ASTM. The testing methods, specifications, and definitions for plastics developed by Committee D-20 now fill two volumes of ASTM Standards (parts 26 and 27) and these volumes not only are the "bibles" of the American plastics industry but they have become international references.

Since ASTM was one of the founding members of the American Standards Association, it has followed through the years that ASTM Standards, upon recommen-dation and acceptance, also become American Standards. As American Standards are accorded both national and international recognition, ASTM has cooperated wholeheartedly with ASA and, more particularly, Committee D-20 on Plastics has taken actions which have led to the adoption of many of its standards as American Standards. In addition, some standards for plastics products have been initiated directly by Sectional Committees of ASA, with these including such examples as B72 on Plastic Pipe, C59 on Electrical Insulating Materials, Z26 on Safety Glazing Material, and Z97 on Safety Requirements for Architectural Glazing. In the future, ASA may play an even larger role in United States and international standardization since steps now are being taken to formally create a national standards body under federal charter.

Another active force in standardization for plastics is The Society of the Plastics Industry. So far, 29 standards for plastics products, formulated by SPI committees working in cooperation with the Office of Product Standards, National Bureau of Standards, U. S. Department of Commerce, have been promulgated as Commercial Standards, and 18 more are currently in progress. In addition, SPI committees are cooperating with other agencies and regulatory bodies such as the Bureau of Explosives and the Interstate Commerce Commission with regard to regulations for plastic containers; with the Building Officials Conference of America, the International Conference of Building Officials, the Southern Building Code Congress, and others with regard to standards for plastics in building construction; and with yet other federal, state, and municipal authorities on regulations governing public safety, health, weights and measures, transportation, labeling, and many other rules applying to the manufacture and use of plastics and the products made of them.

Other standards governing the use of plastics for electrical insulation have been and are being developed and promulgated by the Institute of Electrical and Electronics Engineers, the Insulated Power Cable Engineers Association, the International Electrotechnical Commission, the International Municipal Signal Association, the National Electrical Manufacturers Association, and the Rural Electrification Administration of the Department of Agriculture, to name a few.

For the automotive industry, the Society of Automotive Engineers, both alone and in conjunction with ASTM, issues standards governing plastics used by this industry. The Technical Association of the Pulp and Paper Industry has concerned itself increasingly with standards for plastics used in packaging and printing. The National Sanitation Foundation initiates and administers standards for plastic pipe, and the National Board of Fire Underwriters and its associated Underwriters Laboratories have evaluated plastics and set standards for them for many years.

International standardization of plastics accompanied the geographic expansion of the industry. In 1951, Technical Committee 61 of the International Organization for Standardization was organized to promote international standardization of nomenclature, methods of test, and specifications applicable to materials and products in the field of plastics. ISO/TC61 Plastics now includes 41 countries in its membership, and in 15 years it has done much to reconcile differences between the plastics standards of its member countries. The purpose of this activity is to facilitate "international exchange of goods and services and to develop mutual cooperation in the sphere of intellectual, scientific, technological, and economic activity."

In addition to all of the foregoing, the General Services Administration and the Department of Defense, particularly, maintain continuing programs of standardization specific to the needs of the federal government. Many of the major private consumers of plastics do likewise, either independently or in conjunction with recognized technical societies.

Bibliography

American Gum Importers Assn.   *Natural Resins,* 1936.
Bakelite Review   Forty Years of Progress, Bakelite Corp. Jan. 1950.
Bakelite Review   Silver Anniversary Number, Bakelite Corp. vol. 7, no. 3, 1935.
Bakelite Review   All issues vol. 1, 1929, to date.
Beach, Norman E.   *Plastic Laminate Materials.* Foster Pub. Co., Long Beach, Cal., 1967.
Boles, A. S.   *Industrial History of the United States.* L. Stebbins, Hartford, Conn., 1878.
Boonton Molding Company   *Molded Bakelite,* 1926.
Brauns, F. E.   *The Chemistry of Lignin.* Academic Press, New York, 1952.
Bucks County Historical Society   *Bulletin.* Doylestown, Pa.
The Burroughs Company   *Complete Improved Molding Equipment,* 1926.
Celanese Corporation of America, The Founders and Early Years.   Harold Blancke, 1952.
Chemical Week "1918–1968: The Rise of the United States Chemical Industry," November 16, 1968.
Clement, Luther A.   *A Review.* Rohm and Haas, 1968.
Delmonte, John.   *Plastics Molding.* Wiley, New York, 1952.
Delmonte, John.   *Plastics Engineering.* Penton, Cleveland, 1940.
Dickinson, Thomas A.   *Plastics Dictionary.* Pitman, New York, 1948.
Dow Chemical Company   *Polystyrene, First Twenty-Five Years,* 1963.
DuBois, J. H.   *Plastics.* American Technical Society, Chicago, 1942.
DuBois, J. H., and Pribble, W. I.   *Plastics Mold Engineering.* American Technical Society, Chicago, 1946.
E. I. Du Pont de Nemours & Co., Inc.   *Plastics Department, History to 1942.*
E. I. Du Pont de Nemours & Co., Inc.   *The Story of Cellophane,* 1950.
E. I. Du Pont de Nemours & Co., Inc.   *Paths of Discovery,* 1963.
E. I. Du Pont de Nemours & Co., Inc.   *Nylon, The First 25 Years,* 1963.
E. I. Du Pont de Nemours & Co., Inc.   *The Wide World of Teflon,* 1963.
Durant, Wm. H.   "Comb Industry," *Leominster Enterprise,* Aug. 24, 1923.
Dyle, Bernard W.   *Comb Making in America.* Perry Walton, Boston, 1925.
Eastman Kodak Company.   *Eastman Observes 25 Years of Tenite Plastics,* 1957.
Eastman Kodak Company.   *Kodak Milestones,* November 1953.
Emerson, W. A.   *Leominster, Historical and Picturesque.* Lithotype Publishing Company, Gardner, Mass., 1888.

Gidvani, B. S. *Shellac and Other Natural Resins.* Plastics Institute, London, 1954.

Goodyear, Charles. *Gum Elastic.* McLaren and Sons Ltd., London, 1937.

Griff, Allen L. *Plastics Extrusion Technology.* Reinhold, New York, 1967.

Hall, C. R. *History of American Industrial Science,* 1954.

Haynes, Williams. *Cellulose, The Chemical That Grows.* Doubleday, New York, 1953.

Hercules Powder Company. *A Brief History.* November, 1962.

Hicks, J. S. *Low Pressure Laminating of Plastics.* Reinhold, New York, 1947.

Hicks, Edward. *Shellac.* Chemical Publishing Co., New York, 1961.

Hull, S. M., and Lynn, A. M. "Like Mother Used to Bake," *Western Electric News,* April 1930.

The Hydraulic Press Manufacturing Company. *It Started with an Apple.* 1952.

The Hydraulic Press Manufacturing Company. *HPM Presses,* vol. 2 and 3.

Jacobi, H. R. "The Historical Development of Plastics Processing Techniques," *Kunststoffe,* vol. 55, March 1955.

Jacobi, H. R. *Screw Extrusion of Plastics: Fundamentals, Theory.* Gordon and Breach, New York, 1963.

Journal of Industrial and Engineering Chemistry, vol. 6, no. 2, 5, 7, 1914.

Kaufman, M. *The First Century of Plastics.* Plastics Institute, London, 1964.

Kaye, S. L. *The Production and Properties of Plastics.* International Textbook Company, Scranton, Pa., 1947.

Leominster, *see* Emerson.

Lubin, George. *Handbook of Fiberglass and Advanced Plastics Compositions.* Van Nostrand, Reinhold, New York, 1969.

Macht, Paine, and Rahm. "Injection Molding," *Industrial Engineering and Chemistry,* May 1941.

Mansperger, D. E., and Pepper, C. W. *Plastics Problems and Processes.* International Textbook Company, Scranton, Pa., 1942.

McCann, Hiram. *The Formica Story.* The Formica Company, Cincinnati, 1953.

Mills, Elmer E., Corporation. *Injection Molded and Extruded Plastics.* 1943.

Mumford, John Kimberly. *The Story of Bakelite.* Robt. L. Stillson Co., New York, 1924.

New York Times. "Frangible Bullet," March 15, 1945.

Oleesky, S. S., and Mohr, J. G. *Handbook of Reinforced Plastics of the SPI.* Reinhold, New York, 1964.

"Plastics Molding as Used in Telephone Art," *Western Electric News,* April 1930.

Preiswerk, Dr. E. W. *Invention of Araldite.* Technica, 1965.

Rahm, L. F. *Plastics Molding.* McGraw-Hill, New York, 1933.

Randall, H. D. "Molded Products That Replace Metals," *American Machinists,* October 1929.

Richardson, H. M., and Wilson, J. W. *Fundamentals of Plastics.* McGraw-Hall, New York, 1946.

Rinhart, Floyd, and Marion. *American Miniature Case Art.* A. S. Barnes & Co., New York, 1969.

Saechtling, Hansjurgen. *Hochpolymere Organische Naturstoffe.* F. Vieweg, Braunschweig, 1935.

Sasso, John. *Plastics for Industrial Use.* McGraw-Hill, New York, 1942.

Schack, William. *Manual of Plastics and Resins in Encyclopedia Form.* Chemical Publishing Company, New York, 1950.

Shaw Insulator Co. *Condensite and Other Molding Compounds,* 1920.

Simonds, H. R., and Ellis, Carlton. *Handbook of Plastics.* Van Nostrand, New York, 1943.

Simpson, David C. *Modern Trends in Biomechanics.* Page Bros. Ltd., London.

Smithsonian Institution. *Collection of Heating and Lighting Utensils in The U. S. National Museum,* 1928.

Smith, W. Mayo. *Manufacture of Plastics.* Reinhold, New York, 1964.

Society of Plastics Engineers Inc. *A Guide to Literature and Patents Concerning PVC Technology.* SPE, 1962.

Thomas, Islyn. *Injection Molding of Plastics.* Reinhold, New York, 1947.

Worden, Edw. C. *Nitrocellulose Industry.* Van Nostrand, New York, 1911.

Index

Bakelite Travelcade, 191
Baldwin, Lloyd, 149
Band–Aids, 183
Banks, R.L., 307
Barkdoll, 317
Barnhart, Dr. W., 315
Bateholts, 121, 135
Bathroom, 403
Bauman, 280
Beach, W.I., 388, 425
Bean, C. Thomas, Jr., 395
Beck, J., 390
Becker, T., 268
Beckton Dickinson, 249
Beetle, 159
Belden Mfg. Co., 179
Bell Telephone Labs, 308
Belnor Corp., 288
Bender, Dr. H.L., 425
Bernardinelli, F., 317
Bernhart, 328
Berthelot, M., 290
Bewley, 44, 324
Biaxial orientation, 331
Bill Publications, 419
Birtman Electric Co., 195
Bishop, Dr. P.W., 53
Bitumen plastics, 29
Black, Dr. Otis D., 389
Bliss Co., E.W., 273
Blister packaging, 298
Block press, 327
Blood glues, 6
Blow molding, 346

Blown film, 336
Blown vinyl, 364
Bolton, E.R., 302
Boonton Molding Co., 94, 360
Boonton Rubber Co., 93
Boonton Rubber Mfg. Co., 85, 156
Borer, injection blow, 360
Borg Warner, 253
Borkland, Gustave, 249
Boron fibres, 402
Bothelf, Dr. A., 81
Bottle lining, 355
Boyer, Dr. R.F., 425
Bradley Container Co., 241
Brama, Joseph, 322
Brandenberger, Dr. J.E., 278
Brannon, 179
Braund, J.J., 247
Breaker strip, 382
Breskin and Charlton, 416
Breskin, Charles, 416
Breskin Communications, 421
Bridgeman, Percy, 305
Brillhart, Arnold, 296
British Celanese Ltd., 268
British Xylonite Co., 52
Brock, Frank P., 102
Brockway Glass Co., 242
Brown, Alan, 104
Brown, F.E., 241
Brown, Dr. Frank, 317
Brown, Gordon, 104, 391
Brown, Kirk, 100, 104
Brown, Sanford, 104, 171

DATE DUE

DEC 10 90			
GAYLORD			PRINTED IN U S A